ÉANNA NÍ LAMHNA

THE GREAT IRISH BIODIVERSITY BOOK

ILLUSTRATED BY
Barry Falls

GILL BOOKS

The Author

ÉANNA NÍ LAMHNA has spent all her life looking at wildlife, talking about it on radio and television, writing about it in books and newspapers, and teaching children and grown-ups about it too. She is a native of Co. Louth but has been living in Dublin for most of her life.

I would like to dedicate this book to my grandchildren – Archie, Shay, Hugo, Alice, Teddy, Charles and Tilly. I hope you will get to visit the many wildlife habitats described here, and that all these plants and animals, and the places they live in, will still be there for you to show to your grandchildren.

The Illustrator

BARRY FALLS grew up in rural Northern Ireland, where he spent a lot of time drawing pictures and writing stories to go with them. He works as a commercial illustrator and has won multiple awards. His work has appeared in magazines, books, exhibitions, on textiles and homeware, and in advertising campaigns around the world.

CONTENTS

Introduction

In Ireland we have hundreds of species of wild plants, fungi and birds, and thousands of species of insects. Then there are all the trees, and the mammals, fish and creepy-crawly creatures that live in the sea and in lakes.

But all these plants and animals can only live in places, called habitats, where conditions are just right for that particular species. Creatures that live in woodlands are different to those in ponds, and what lives on a bog couldn't survive on the seashore. This is called biodiversity, and the rules that determine what lives where are always the same.

All the food on earth is made by the plants, the animals eat the food, and the decomposers break the dead plants and animals down again. Every single habitat has these three groups of living things.

PLANTS make food from the air – or at least from the carbon dioxide in the air. Their leaves are green because they contain very large green molecules called chlorophyll. These molecules can break open the carbon dioxide in the air, using sunlight for energy. The carbon is saved in the plant as sugars, and the oxygen part of the carbon dioxide is given off for us all to breathe.

This process is called photosynthesis, and without it no life would be possible on Earth. This is the only planet in the universe where this happens as far as anyone knows ... so far, anyway. The food contained in the plants is eaten by animals. Animals that rely on just plants for food are called **HERBIVORES**. Some eat the whole plant. Others might only eat the leaves or the roots or the nectar produced in the flowers. But they are all herbivores and can only digest plants. Rabbits and pigeons can't actually digest any meat, even if they eat it by accident. Every habitat has herbivores living there, feeding on the plants.

CARNIVORES can only eat meat – that is, the flesh of other animals. They cannot digest any plants at all. Otters, eagles and frogs are all carnivores, as are ladybirds, swallows and sharks. Each habitat has its own carnivores. Some carnivores can eat other carnivores – there is a food chain of who eats whom. Sparrow hawks eat robins that eat spiders which eat flies.

Then there are **OMNIVORES**. These animals can digest both plants and animals, so they have a bigger choice of food to eat. Humans are omnivores, as we have the enzymes to digest both meat and plants. Even people who choose never to eat meat are still omnivores. Animals that are omnivores can be very successful, as they can eat a different food if one sort becomes scarce.

DECOMPOSERS break down all the dead plants and animals. Animals like woodlice or millipedes can be decomposers, in the first instance, but the main work is done by bacteria and fungi. Life on Earth wouldn't last long without them to keep things moving.

You will meet all these groups of living things in each of the six habitats described in *The Great Irish Book of Biodiversity*.

Some of the plants and animals described in this book are much rarer than others. You could find a few of them if you are really interested and know where to look, but others are once-in-a-lifetime sights.

Keep an eye out for the five Great Irish Nature Challenge pages in this book to find out more about these rare and wonderful creatures!

Éanna Ní Lamhna

Woodlands

WHAT IS A Woodland?

A WOODLAND IS a place with lots of trees, obviously. But does just having trees make a place a woodland?

If I plant a large field with trees some weekend, is that then a woodland? No, it's not – it's just a crop of trees in a field. To be a real, functioning woodland there have to be lots of other things there too. There must be woodland flowers and ferns and mosses. And, of course, there must be woodland creatures that live there and nowhere else, such as particular types of birds and mammals and insects and other creepy-crawlies. There must be tall canopy trees that reach up to the sky and use a lot of sunlight in order to grow. Then, underneath, there must be smaller trees and large shrubs, just waiting for their chance to shoot up as soon as one of the canopy trees falls.

IS IT MORE THAN JUST TREES?

IT MOST CERTAINLY is. Trees are just the start. Woodlands have special plants that flower early when there are no leaves on the trees and the light gets right down to the forest floor. Then, when the leaves have come, in April and May, those plants on the forest floor that can live in low-light conditions, such as mosses and ferns, make a strong appearance.

Insects that feed on green leaves then begin to appear in the trees. In turn, they will be food for the birds who are specialists in finding them. Other animals depend on the nuts and berries produced in autumn. It will be a long time before the field of trees becomes such a woodland – but it will eventually.

Trees, plants and flowers

Trees are the most important woodland plants. Long-established woodlands have great mature trees, new young saplings and ancient trees with cracks and holes. They have dead trees on the ground that have come to the end of their lives but still provide a home for creatures who live in dead wood.

Oak woodlands

Oak woodlands, where the big canopy trees are oak trees, are common native Irish woodlands. Often **BEECH** trees have been planted among them. The soil is the sort of acid soil that suits oak very well, and also the **BIRCH** and holly trees that grow underneath the main canopy. **HOLLY** trees are evergreen, with leaves all year round, and they grow really well from October to May. This is because the oak, beech and birch, which are deciduous, have lost their leaves and there is more light in the woodland.

Other sorts of woodlands

In places where the soil is not so acidic, woodlands will have **ASH**, **WILD CHERRY** and even **ELM** as the big canopy trees. The smaller trees will be **HAZEL** and **HAWTHORN**. If the soil is really wet, trees that can put up with wet conditions, such as **ALDER** and **WILLOW**, will grow.

Does ivy deserve its bad reputation?

Ivy loves to grow in woodlands. It has really dark five-pointed leaves that manage well in the low light in summer on the forest floor. But in order to make flowers and berries it has to make different shaped leaves, and it needs stronger light for this. It climbs upwards – up tree trunks, dead trees, any walls or poles that might be nearby – desperate for more light. It does no harm to living trees. The flowers it produces in October provide lots of nectar for bees, and its black berries in February are food for hungry birds.

Woodland flowers

Woodland flowers appear early in the year. Many of these are white so that the insects looking for pollen and nectar can find them. **WILD GARLIC**, **ARUM LILIES**, **WOOD ANEMONES**, **SORREL** and **ENCHANTER'S NIGHTSHADE** are all white. But yellow works too, and **PRIMROSES** and **LESSER CELANDINES** often occur in large numbers.

These flowers all send up their leaves first, and when the plant has grown big and strong enough, then the flowers form. **BLUEBELLS** do it the other way round. They have fat bulbs under the ground, and the flowers come straight up from those. Afterwards, the leaves last a long time – fattening up the bulbs again for next year's flowers.

The native Irish bluebell

Native bluebells are becoming much scarcer, as they can interbreed in the wild with the garden bluebell called the **SPANISH BLUEBELL**. The native species provides early nectar to important pollinators like bees and butterflies, aiding in the pollination of other plants as well. So never plant Spanish bluebells in the garden.

Woodland insects

Many insects that live on leaves are coloured green so that their enemies won't see them. There can be thousands of **GREENFLY** on the leaves of an oak tree, all sucking away at the juices in the leaves through long needle-like mouthparts.

Moths

Moths like nectar too. Honeysuckle flowers put out a lovely perfume – only at night – to attract moths to come and sip their nectar. There is not enough nectar in just one flower for a hungry moth, so they visit several, carrying pollen on their heads from one flower to another and pollinating them – which of course was the honeysuckle's intention all along.

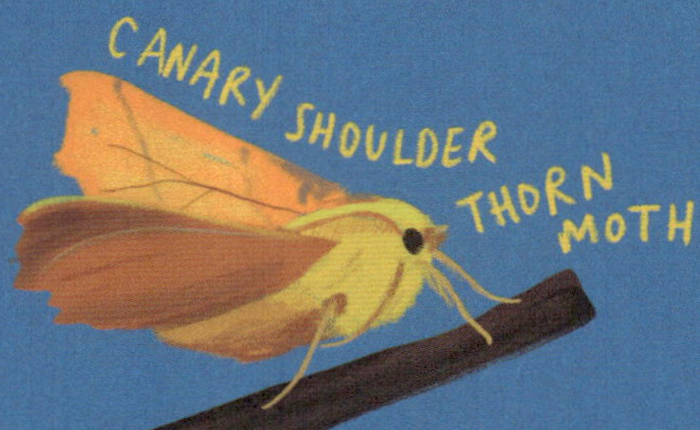

Avoiding being eaten

The **CATERPILLARS** of woodland butterflies and moths devour leaves as well. While some are coloured green to blend in with the leaves, others are covered in long prickly spines that hurt birds' mouths. Some caterpillars have great camouflage and don't look like birds' meals at all. They can look like a twig or a snake or a bottle brush.

Bees

Bumblebees are herbivores. In spring they collect pollen from the catkins of hazel and willow trees and bring it back to their nests to feed the baby bees. They visit the woodland flowers to collect the nectar they produce and bring it back to their nests to make honey for food for themselves.

Ladybirds

Ladybirds like to eat greenflies. They are also brightly coloured to warn birds not to eat them because they taste terrible. Typical woodland ladybirds are the **11-SPOT**, which is red with black spots, the **CREAM-SPOT**, which is maroon with 14 spots, and the **ORANGE LADYBIRD**, which has 12 to 16 cream spots.

Shield bugs

Shield bugs like to eat leaves. They walk over the leaves first, putting a horrible smell on them so the greenflies won't eat them. This is why they are also called **STINK BUGS**.

Wasps

Wasps live in big nests made out of paper. They make this paper from chewed-up wood, so they often have nests in holes in trees in woodlands. The worker wasps collect thousands of **APHIDS** and other small insects, chew them up and bring them back to the nest to feed the young grubs. But grown-up wasps have a sweet tooth, and they only like sweet things.

In autumn, wasps feed on nectar and fallen fruit. But, usually, they get their sweet food in the nest from the baby wasps – the **GRUBS**. When the grubs are fed with chewed-up insects they produce a sweet dribble that the adults lick. Yum yum!

Woodland animals

While animals from other habitats, such as gardens or grasslands, might occasionally be seen in a woodland, there are some for whom the woodland is their home. They need the trees for food, and in the case of squirrels and pine martens, the trees are where they build homes for themselves, high up among the branches.

Bats

Bats come out at night and fly through the air, hunting and eating other animals that fly around at night, such as moths, midges and mosquitoes. They can wheel and turn and fly very fast through the trees, never crashing into anything. Our biggest bat – **LEISLER'S BAT** – hunts in woodlands, as do our three species of **PIPISTRELLE** – the soprano, the common and Nathusius's.

Bats are not blind, but seeing at night without light is not easy. They use **ECHOLOCATION** to be sure not to crash into anything. They emit high-pitched sounds that bounce off trees and moths and the other insects they are hunting. This is how they know exactly where everything is.

Pine martens

Pine martens are lovely chestnut-brown creatures that are really good at climbing trees. Their Irish name is *cat crann*, which means 'tree cat'. Pine martens are the natural enemy of squirrels, but they find it quite hard to catch **RED SQUIRRELS**, which are much lighter than them – they can run out to very small branches at the tops of the trees that would break under the weight of a pine marten.

But pine martens are well able to catch the heavier **GREY SQUIRRELS**, which often feed on the ground too. Pine martens particularly love cherries – as a dessert, I suppose – and feast on those found on the wild cherry trees in July.

Badgers

Badgers' homes are underground tunnels called setts. A badger family can live in the same sett for hundreds of years, and big ones have several entrances. While badgers are the only animals that can eat hedgehogs, and they visit gardens to dig in the lawn for earthworms, they are very fond of elderberries, mushrooms and blackberries too, which they find in the woods.

Deer

Deer are very fond of eating young growing trees, so they love woodlands. In places where there are too many deer, they eat all the young trees. This means that eventually the woodland will cease to exist because there will be no new trees to take the place of the older ones when they die. Wolves are deer's natural enemy, but we no longer have wolves in Ireland so the deer can do what they like.

Squirrels

Squirrels depend on trees such as hazel and oak for nuts. They collect these and hide them in special stores to eat in winter. They don't **HIBERNATE**, so they need this winter food. In spring and summer, they often eat the bark and buds of the trees and can cause harm to them.

Woodland birds

Woodlands are the best habitat of all for Irish birds, as there is so much food and shelter there for them all year round. They can't see each other because of the trees, so woodland birds are very good singers, as they communicate by sound.

Deadly flying killers

LONG-EARED OWLS hunt at night in woodlands. They pounce silently on wood mice and rats who never hear them coming, as they have feathers on their legs right down to their toes.

TREECREEPERS can be found on oak trees with rough bark. They walk up the trees holding on with their strong clawed feet and balancing on their tail feathers. They poke their thin curved beaks into the tree bark and eat the insects and caterpillars that live there.

Our smallest bird, the **GOLDCREST**, prefers evergreen trees, where it builds its nest from spiderwebs – having gobbled up the resident spiders first.

The **WOODCOCK** can be very hard to see. This is a wader with a long beak that sits tight on the woodland floor. Its feathers are exactly the same colour as the leaf litter it is sitting on. It feeds on earthworms and insects in the ground. Its eyes are so far back on its head that it can see what is going on behind it – just like teachers.

Milk-producing birds

Some woodland birds live entirely on plants. The **WOOD PIGEON** is one of these. It is particularly fond of the berries of elder trees and hawthorn trees. It does not depend on insects to feed its young.

It produces a cheesy type of milk in its crop, which is a sort of flexible bag in its chest. It regurgitates, or spits back up, the milk when the bag is full, and the young birds drink this. This means it can have a nest at any time of the year, not just in summer.

Clever birds

JAYS are the colourful member of the **CROW** family, with brilliant blue wing feathers. Like all crows, they eat birds' eggs, slugs and small mammals, but in autumn acorns are their treat. They collect all they can and bury them one by one for eating later in winter, when there's less choice. Any they don't eat grow into oak trees, so jays are important planters of oak woods.

LONG-TAILED TITS fly in small flocks. They look like pink flying spoons. In summer they eat insects, caterpillars and spiders, but when times are hard in winter, they eat seeds and buds and are glad to get them.

The **GREAT SPOTTED WOODPECKER**, which only returned to breed in Ireland in 2008, now can be found in every county. It cracks open rotten timber with its sharp, long bill and eats the larvae of moths, wood wasps and other wood-eating creatures. It can also eat seeds and nuts when it comes across them.

Woodland decomposers

All sorts of leaves and other dead plant matter end up on the forest floor. Why after, say, five years of this are you not up to your neck in dead leaves as you walk through the woods? Because they rot away, of course. Rotting or breaking down is done by the decomposers.

Deciduous and evergreen trees

The woodland floor is covered in dead leaves by the end of November. The deciduous trees, like **OAK**, **BEECH** and **BIRCH**, lose all their leaves together once cold autumn comes. They have to, as they have big soft leaves that would freeze solid in the cold winter and then the trees would die. Evergreen trees, such as **HOLLY**, have a thick waxy covering on their leaves. **SCOTS PINES** have long needles with a sort of anti-freeze solution inside that stops them freezing. But even evergreens have to grow new leaves too. They do it in small doses all year round, not in one fell swoop like the deciduous trees.

Fungi

Fungi are expert decomposers. They send a complete network of long thin white threads through the soil, the dead leaves, the fallen trees or those upright trees with dead patches in them – like a wood-wide web of strands. These break down the dead leaves and use up the carbon to make more white threads. Occasionally they send up **MUSHROOMS**, which can make **SPORES**. This allows them to spread when the spores are scattered. In autumn, the woods are full of these mushrooms – on the ground, on the fallen trees and even on cracks in living trees. Mushrooms can be white or brown or red or violet or olive coloured, but they are never green like leaves.

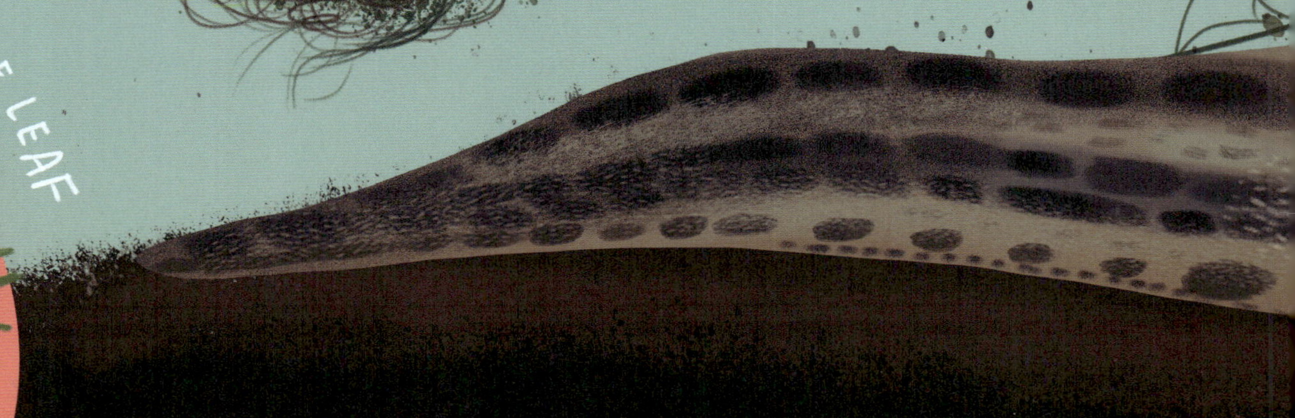

The rest of the decomposer gang

Lots of animals help with this breakdown work as well. **WOODLICE** eat the dead leaves and get whatever nourishment they can from them before pooing out the undigestible bits for the fungi to finish off. Woodlice all have 14 legs. The **PILL BUG** is a woodlouse that can roll itself into a tiny black ball.

MILLIPEDES also help to break down dead leaves and leaf litter on the forest floor. Although they are called millipedes, they don't have a thousand legs – more like 200, at the most. Some of these can curl into a ball too, like the **PILL MILLIPEDE**. The **SNAKE MILLIPEDE** glides along like a tiny snake and coils up like a spring when disturbed. This is to pretend to hungry black beetles that they are not creatures at all and so not worth eating. The **GREAT GREY SLUG** eats dead plants – unlike its garden cousins, which only eat fresh, green growing plants. So it is also a member of the decomposer gang.

A woodland walk

If you want to see exciting things on a woodland walk, there are a few things to remember. No headphones – otherwise you won't be able to hear the woodland sounds. No talking – the wildlife will hear you coming and clear off immediately. Better not bring the dog either!

Leaves and flowers

See how many different kinds of trees you can notice. Better still, collect leaves fallen on the woodland floor and see how many types you find. Woodland flowers – conveniently – are more common near paths because there is more light.

See if you can find white ones, yellow ones, pink ones, purple ones, blue ones. You won't find green ones – that is the colour of leaves – or red ones because pollinating insects such as bees cannot see the colour red and so never visit them to pollinate them.

Footprints and feathers

Can you find traces of what has been there before? There might be animal or bird footprints in wet, muddy parts – deer, perhaps, or a badger or just someone's dog. Look at the size of the print and the number of toes – each species has a different footprint. Take a photo to check it out later. Feathers are often to be found, as birds' feathers wear out from flying, so they drop them and grow new ones.

Grey ones could be from **PIGEONS**, black ones from **CROWS**. But you could also find a blue **JAY** feather, or a lovely brown dappled one from a **WOODCOCK**, or a copper one from a **PHEASANT**.

Droppings

Scientists are interested in animal droppings too. A **PINE MARTEN** that has been gorging itself on wild cherries will produce a dropping full of cherry stones. **BADGERS** are very clean animals and use a special place called a latrine to deposit their poo. You might come across this, with elder trees growing from the droppings of a badger that devoured elderberries a few years ago. **LONG-EARED OWLS** and **KESTRELS** swallow their food – mice and rats – whole. Later, they cough up all the bones and fur they couldn't digest in a pellet – often under a tree where they have been roosting. If you find one of those you can take it home and tease it out to see all the little bones of their meal.

Leftovers

Food remains tell a story too. Nutshells split in half and left in a hole in a tree were eaten by **FIELD MICE**. **SQUIRRELS** leave jagged edges on nut shells. Nuts wedged in tree bark were poked in by **MAGPIES** or **WOODPECKERS** to make them easier to crack open.

Use all your senses

Stop and close your eyes all the better to hear bird song or the drumming of a woodpecker. Use your nose to detect the musky smell a **FOX** leaves behind – or the earthy smell of **MUSHROOMS** in autumn.

Woodland walks are great!

THE GREAT IRISH NATURE CHALLENGE

Rhinoceros beetle

This beetle looks like a small version of the huge rhinoceros beetle from the tropics, and it's found in some Irish woodlands, particularly mature oak and beech woodlands. They walk over fallen dead trees, and when they get to a particularly soft place, the female bores a tunnel in the rotten wood while the male stands guard. The eggs are laid in the tunnel and the larvae live in there for three years, eating all round them. When they emerge as adults, they can fly using their back pair of wings. They are active both day and night, so you could try looking for them with a torch in the dark to make the expedition even more exciting.

The goosander

The goosander is the only native Irish duck to nest in a tree. The female nests in a hole in a tree, which she lines with rotten wood, leaves and feathers to make it comfortable. There, where she feels safe from danger, she lays up to 13 eggs and hatches out the ducklings after sitting on them for five weeks. She can't feed them up in the tree, so when they have all hatched, they have to jump out of the nest. They can't fly yet, so they bounce down to the ground – they don't get hurt because they are so light. Then they run down to the lake with their mother, where they learn to dive for fish. The oak woodlands in Glendalough, between April and June, are a good place to look for them.

Remember, these plants and animals are much rarer than others – how many can you find?

The stinkhorn

This stinky mushroom smells like rotten meat or a burst sewer. It produces this smell to attract flies and slugs, who are fooled into thinking a meal of rotten meat is waiting for them. The smell comes from the olive-green slime that drips off the cap. This slime contains the spores, so when the flies and slugs eat it and move away, they spread the spores and new smelly fungi can grow. Walk through a woodland in August and September, sniffing as you go, and you might find this 25-centimetre-high fungus that grows on buried dead wood. It loses the horrible smell once the slimy cap has been licked off.

The Kerry slug

The Houdini of the undergrowth, this slug can squeeze through the narrowest of cracks. It is one of the rarest slugs in Ireland and is protected by law, so you can only look at it. You can't collect it and bring it home – which is just as well, as this slug can squeeze itself so flat, it can slide through the narrowest of cracks, such as out of a closed matchbox. It has only been found, so far, in Kerry, south-west Cork and County Galway, but it could be in other woodlands too. Look for it in deciduous oak woodlands, or try coniferous woodlands. Wet, cloudy days are the best times to find it. It feeds on the lichens and algae that grow on trees. It does no harm at all to growing trees. Take a picture to prove that you have found it.

Boglands

WHAT IS A BOG?

THE WORD *BOG* comes from the Irish language, and it means a soft, wet, squishy place where the ground is made of peat or turf. Ireland has always had bogs because it rains a lot here. There are two different types of bogs, depending on how they were made long ago.

RAISED BOG

THE RAISED BOG is the oldest type. These were made when large shallow lakes, left behind when the last ice age finished, about 10,000 years ago, began to fill up with plants that grew around the edges. These plants grew every year, and when they died back, they sank into the lake water. After a long time, the original lakes were all filled up with the dead plants. These dead plants weren't able to rot away because they had died in such a wet, watery place, which had no decomposers.

New sorts of plants then grew on top of them, and these plants could only get minerals from the rain – and very few minerals at that. But one sort of plant was good at this and grew really well. This is a moss called SPHAGNUM MOSS. Not only can it grow where there are few minerals, but it also loves to grow in very wet places too, as it can hold twenty times its own weight in water. So it grew and grew and built the bog right up over the surrounding countryside – which is why it was called a raised bog.

BLANKET BOG

THE OTHER SORT of bog – blanket bog – occurs on our mountains, which have the highest rainfall. Sphagnum moss grows here too and forms a blanket over the land. These bogs began to form 4,000 years ago, when our weather became much wetter than it had been before.

Bog plants

The most important bog plant is the one that made the bog happen in the first place – sphagnum moss. There are 25 different species of sphagnum, in every shade of green through pink, red and purple, bronze, brown, orange and yellow, even drying out to white if there is a continuous dry spell.

Bog flowers

BOG COTTON looks like pieces of white cotton wool on stalks, but it is not strong enough to be woven into cloth like real cotton. **LOUSEWORT** has purple flowers and was once (wrongly) thought to encourage lice in sheep. **BOG ASPHODEL** has a lovely golden-yellow flower. Some orchids even grow on bogs, such as the tiny **BOG ORCHID** with yellow-green flowers that grows among the sphagnum and the **LESSER TWAYBLADE**, which has only two large leaves.

Insect-eating plants

SUNDEWS are small red plants with round leaves covered in tentacles. These tentacles produce an extremely sticky glue. When an insect lands on the leaf it gets stuck, and all the tentacles bend towards it and the leaf folds over. The insect dies because all the sticky glue stops it breathing. The sundew then produces strong enzymes that dissolve the poor insect and then the plant absorbs this liquid. Several hours later, job done, the leaf opens up, dries off and the remaining hard bits of the insect blow away. It is ready to trap another one.

Special flowering plants called **BLADDERWORTS** grow in bog pools. These have lots of leaves on which there are tiny bladders, like little bags. If a water insect swims near one of these bladders and touches it, it makes the bladder shoot open and suck in a gulp of water together with the insect. Once inside, the insect cannot get out and it is drowned and digested by the enzymes produced by the plant.

Woody plants on bogs

The bog surface is not all the same. Around the edges can be drier, and it is here that trees such as birch, willow and Scots pine can grow. They can put up with having quite wet roots, as even the edges of bogs can become quite wet when there is lots of rain. **HEATHER** often grows in drier parts of the bog too. Ling heather has small purple flowers, while those of the bell heather are bigger and a pinky purple. Of course, if you find a stand of white heather, you should keep a piece of it, as it brings good luck.

Bog insects

The most noticeable insects on bogs occur up in the air. These flying insects are looking for mates and on a sunny day in summer there can be lots of them about. When mating has taken place, the females have to lay their eggs. Butterflies and moths lay their eggs on plants, while midges and dragonflies lay their eggs in the water.

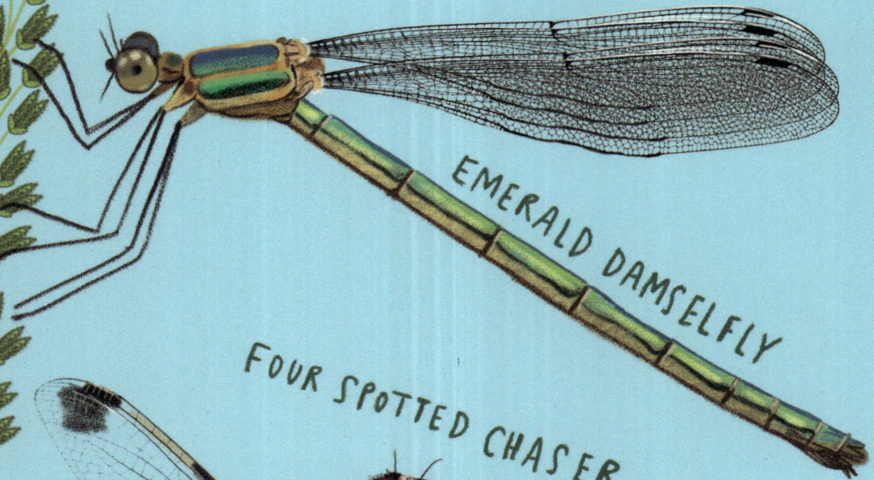

Dragonflies and damselflies

These four-winged insects fly over the bog in summer looking for smaller insects to eat and for other dragonflies with which to mate. Mating males clasp the female by the head and hold on to her. They fly around together like this – sometimes for an hour. Eggs are laid in the bog pools and the male lowers the female into the water and back out again each time.

The **LARVAE** are ferocious carnivores, grabbing any passing creature they meet under the water. Of the 13 species that occur on bogs, the **EMERALD DAMSELFLY** and the **FOUR-SPOTTED CHASER** are the most common.

Butterflies and moths

The wonderful **EMPEROR MOTH**'s caterpillars feed on heather leaves and change colour as they grow from one stage to another. They start out black in early summer and are not noticeable, as there are no flowers yet on the heather. By August they are huge and bright green with black bands, pink spots and very poisonous bristles. The adult moths are brown and white with huge spots on each wing like owls' eyes, which terrify any bird that goes near it.

The caterpillars of the **LARGE HEATH BUTTERFLY** feed on purple moor grass all summer and overwinter as caterpillars, which is unusual. Their bright green cocoon is made the following May, and the adult butterfly emerges in June to mate, lay eggs and die.

The **GREEN HAIRSTREAK** is our only green butterfly. It flies around bogs in May and June laying eggs on plants that the caterpillars will eat, including heather and bilberry (*fraochāns*). The caterpillars feed on the leaves of these plants for 25 days then fall to the ground to pupate. Some books say that the caterpillars can eat each other.

Bog pools

Bog pools are full of insects, all trying to kill and eat each other. **WATER BOATMEN** hold onto plants at the bottom to avoid floating up to the surface and being gobbled up. They eat the tiny algae in the water. But they do have to swim up carefully every so often for air.

BACK SWIMMERS stay still on their backs, just under the water's surface. If a tiny insect falls in, they grab it quickly and eat it. **POND SKATERS** walk across the surface of the water – keeping well away from the back swimmers – and spear **WATER FLEAS** and **MOSQUITO LARVAE** with their needle-like mouthparts.

WATER SCORPIONS hang from the surface by a tube at the end of their body that they breathe through, while using their claw-like front legs to grab any small creatures that swim by.

The **RAFT SPIDER** is the largest spider in Ireland. Its body is 25 millimetres long and its leg span is 70 millimetres. It sits beside bog pools waiting for insects on the water. When it detects one, it darts across the water surface to catch it. It doesn't sink because its broad leg span and special hairs allow it to glide across the water. It eats dragonfly larvae, tadpoles and aquatic insects, but is harmless to humans.

Bog animals

Only animals that can find enough food and shelter to stay there all the time actually live on the bog. Other animals, such as foxes and badgers, will visit to see if they can catch any ground-nesting birds for food. The bog pools are where amphibians can live, as they feed on the water insects there.

Hares

Drier bogs are a hare's favourite place to live. They are herbivores and they eat young heather shoots and the grassy parts of the cotton. They make their nest or **FORM** in sheltered places above ground among the heather. They can sleep with their eyes open so that they are not surprised by foxes or other predators creeping up on them at night. Young hares are called **LEVERETS**.

Newts

Look into a large bog pool and you may very well see the native smooth newt swimming around. Newts are **AMPHIBIANS**, which means that they can survive in water and on land as they have two ways of getting oxygen – through their lungs and directly from the water through their cold, slimy skins. They look like tiny dragons, with four legs and a long tail.

Spring is when mating happens, and the males dress up for the occasion. They develop a red or orange belly and a silvery tail stripe, and they have a big wavy crest all the way down from the head to the tail. The female is much less glamorous and is brownish in colour. The eggs are laid one at a time and wrapped in the leaves of plants growing in the pool. Then the newts leave the pool and spend the rest of the time on land, hibernating in thick vegetation in winter. They are much harder to see then.

Bog birds

Many of the bog birds nest on the ground, as the trees only grow around the drier outside edges. They patrol their territory by keeping an eye on it from the air and singing to keep other competing males away. It can be much harder to keep away a hungry fox or a crow.

Bog birds

Bogs are great places to see MEADOW PIPITS. These small birds nest on the ground and eat insects such as daddy-long-legs and smaller larvae. The MERLIN is its deadly enemy. This bird of prey flies low over the bog, and if it frightens a meadow pipit into flying away, it will fly after it non-stop until the poor meadow pipit drops dead from exhaustion.

CUCKOOS like bogs because they are able to eat the big hairy emperor caterpillars that live there. They lay their eggs in several meadow pipits' nests – one in each – and then the meadow pipit has to rear the much larger baby cuckoo because it hatches out first and throws out the pipit's eggs.

Snipe

Snipe are bog birds that nest on the ground, and their feathers are coloured and striped so they are very well camouflaged among the bog plants. They sit tight if there is a disturbance, only rising to fly away at the very last moment. They make a weird bleating call, not with their voice but by diving fast from a height and letting the wind rush through their tail feathers.

Geese

In winter, a bog is a good place to see the geese that visit here from further north because their feeding grounds are covered with snow.

The **GREENLAND WHITE-FRONTED GOOSE** comes in flocks to the bogs to feed on the nutrient-filled underground roots, or rhizomes, of the bog cotton and other bog plants.

Curlew

Curlew have really long bills with a curve at the end. They nest on the ground and they are well camouflaged by their feathers. They eat insects and the berries that grow on bog bushes. Their call, 'cour-lee', is what has given them their English name.

Red grouse

Red grouse are big hen-like birds that live on bogs and feed entirely on heather. Heather is tough and hard to digest, so grouse have to eat lots and lots to get enough nourishment. And then they produce lots and lots of droppings too.

They make two quite different types of poo: lots of brown pellets from the hard, indigestible heather stalks, and shiny, soft black blobs like chocolate mousse that come when they have finished digesting the softer, more nourishing parts of the heather.

Bog finds

Normally, plants and animals that die decompose or rot away. This happens because bacteria and fungi use the dead plants as food for themselves and break them down. They are often helped by small animals such as woodlice and millipedes, as we saw in the woodland section. But there are no decomposers in bogs!

Why are there no decomposers in bogs?

Bogs are completely waterlogged. The water is right up to the surface, and there is no oxygen at all in this water. Fungi and bacteria are living things, and they need oxygen to break down the dead plants. To make it even more difficult, the whole place is very acidic, and the fungi and bacteria cannot bear that either.

So nothing decomposes or rots away. This year's sphagnum moss dies and falls on top of last year's moss. Next year's moss grows on top of that. If you dug deep down, you would find sphagnum moss that grew hundreds of years ago. Except it wouldn't look like moss at all.

The weight of all the dead moss on top of it would have pressed hard on it and changed it into the dark brown stuff that we call peat or turf.

It is all still there if the bog has never been touched or dug up since it was formed.

And anything that fell into the bog over the centuries will still be there too – it won't have rotted or decomposed away either. You can visit the National Museum of Ireland in Dublin to see what else has been discovered in Irish bogs.

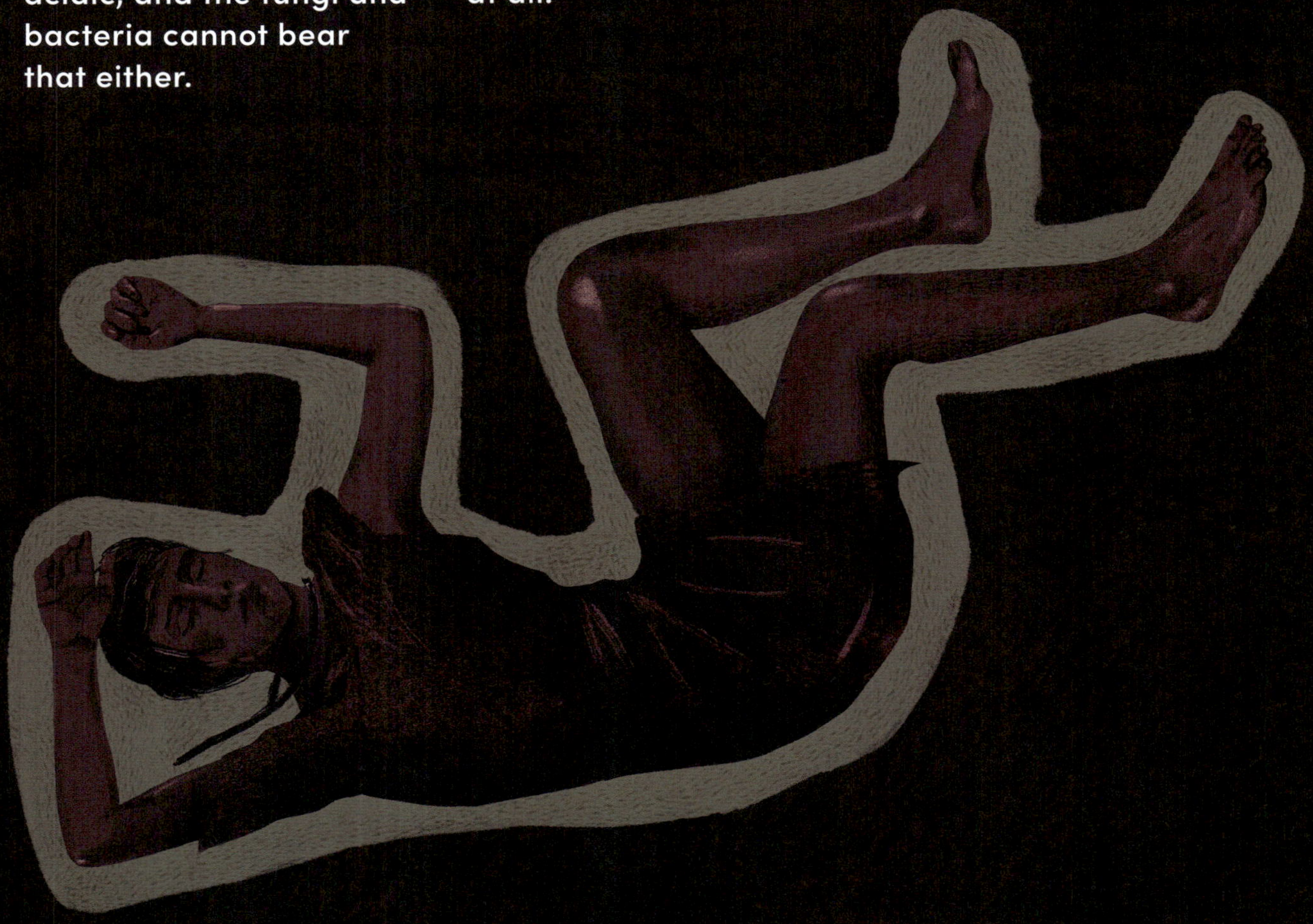

Bog bodies

Bodies of people who were murdered long ago have also been found in bogs. The oldest is **CASHEL MAN**, whose 4,000-year-old body was discovered in Cashel Bog in County Laois in 2011. He had been killed as a human sacrifice. People at that time thought that killing him would please their pagan god and stop the god from doing harm to them or their crops. I wonder did it work?

Bog treasure

People hid treasure in the bog too. Three pieces of Irish **GOLD JEWELLERY** dating to 2300–2000 BC were found in a bog at Coggalbeg, County Roscommon, in 1945. They were part of a necklace belonging to a very rich local ruler. **A BOOK OF PSALMS**, from the Bible, written on parchment in AD 800 by monks at a nearby monastery, was found in Faddan More Bog in County Tipperary. It was probably hidden to keep it safe from the Vikings. The **DERRYNAFLAN CHALICE** was discovered as part of a hoard of altar vessels in Littleton Bog in County Tipperary. It is one of the most spectacular items ever found in a bog in Ireland.

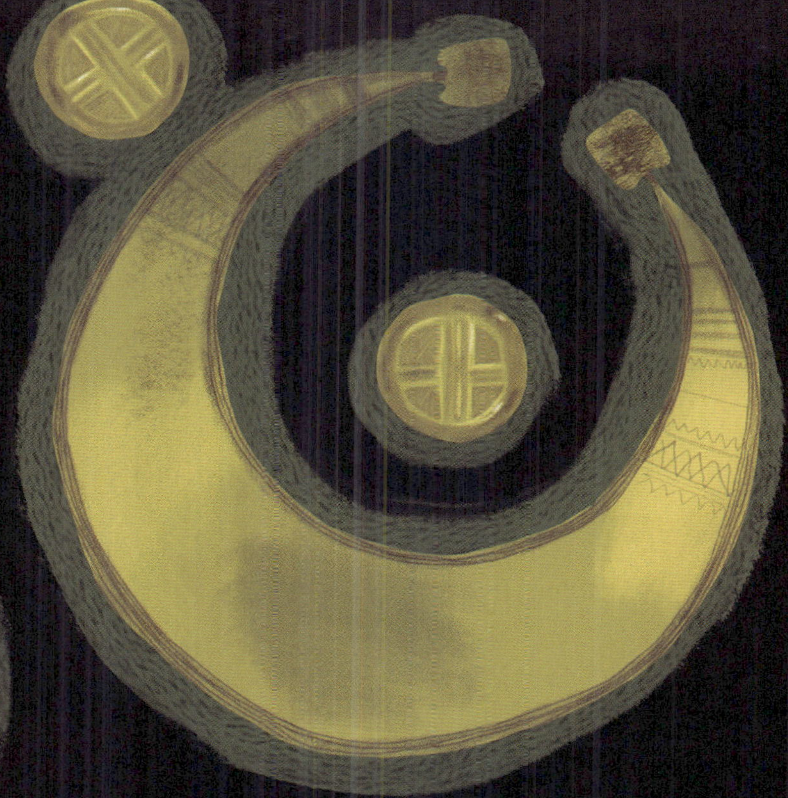

Bog butter

People long ago used to store food in holes in the bog because they had no fridges to keep it fresh. It lasted forever if they forgot to go back for it. A lump of butter found in Knockdrin Bog in County Offaly was 3,600 years old. It probably tasted awful.

A bog walk

Bogs are great open spaces that can be very soft underfoot. If you look carefully at the ground, you will quickly learn that some species of sphagnum grow in drier spots, so these are the ones to put your feet on if you are walking on the bog itself.

Midges

Bogs are great places for a walk, but be careful if you go there on a calm, still day in summer. You are likely to meet clouds of **MIDGES**. These are out for one thing – to find a creature with lots of blood, because the females need this nourishment to be able to lay eggs.

Most of them are content to bite butterflies or dragonflies. However, there are four species that bite us – with tiny little bites that are terribly itchy. Another group don't bite but feed off our sweat, so they make our heads terribly itchy too. Once the wind comes, they are all blown away.

Sounds

Bird songs carry for a long distance. On a sunny day the sky can be full of the song of the **SKYLARK**, but it is very hard to actually see them.

Lichens and fungi

Drier places may be covered in crunchy lichens such as the **DEVIL'S MATCHES LICHEN**, which has scarlet fruiting bodies at the tips of its branches. Look out for fungi too – there aren't many, but some of them have very vivid colours.

They can only grow in the living layer of sphagnum so are quite small. **BOG BEACON** has a bright yellow or orange head. **SCARLET WAX CAPS** are crimson all over.

The **HEATH NAVEL** is an elegant white one growing on vertical peat cuttings.

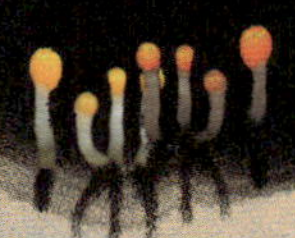

Ghostly lights

As the light fades, look out for the **WILL-O'-THE-WISP**. This is what the strange pale lights hovering over the bog are sometimes called. From a distance, it looks like the light of a candle. It can be still, or it can move. If you get close up, it has a cloudy white appearance, but it is very luminous. Long ago people were afraid to go near it because they thought that the lights were lanterns carried by evil spirits that would lure you into a watery grave. They are actually the flare of burning methane, which is a gas given off by the slowly decomposing turf. This methane is lit by a second gas called diphosphane, which bursts spontaneously into flame when it meets air.

Long dead trees

On blanket bogs, look out for **BOG DEAL**. These are the remains of pine trees that grew here thousands of years ago and were covered by the growing blanket bog when the weather got wetter and colder. They are very dark and hard and have lovely twisted shapes that artists like.

THE GREAT IRISH NATURE CHALLENGE

Crane

The crane is a really large bird, reaching up to a person's shoulder if they were side by side. Cranes spend the winter in southern Europe and migrate north to breed in summer. They nest in wet boggy areas, and we used to have loads of them in Ireland.

When we drained many of the bogs, the cranes didn't bother nesting here any more but kept going to Sweden and Norway. Lately, some of them are breeding here again, and the newly wetted bogs are places you could look for them. You might even catch them doing their wonderful courting dance.

Large marsh grasshopper

This huge green grasshopper can be up to four centimetres long and occurs on Irish bogs, but hardly anywhere else in Europe. It has a streak of red on its upper legs and black knees, so you won't mistake it. It lives only in the very wet parts of the bog, and you should be able to hear it chirping in August and September.

It makes these loud sounds by hitting its hind leg against its forewing – which it does to impress a female. It stops immediately if anyone comes, so you have to creep up on it very carefully. It can hop really far and even fly, too, with its four wings.

Remember, these plants and animals are much rarer than others – how many can you find?

Hen harrier

The hen harrier is a bird of prey that hunts over bogs – not for hens, despite its name, but for small birds, birds' eggs and any rabbits or small hares it can find. The males are pale grey all over with black wing tips. They look completely different to the females, who are bigger than them and are dark brown all over with a stripey tail.

They can nest on the ground or in young conifer plantations, and in summer they hunt over bogs in upland areas. The male catches food and passes it to the female in mid air. She never misses catching it.

Pitcher plant

If you are dying to see a plant eating an insect, then you must go and find a pitcher plant. The Canadian pitcher plant was brought to a bog in County Roscommon in 1906, and since then it has been planted in other bogs too. It is great at catching insects. The pitcher-shaped leaves have lovely nectar that contains narcotics.

The insect that drinks this becomes dizzy, loses its footing on the leaf's slippery edges and slithers down into the watery depths. It quickly drowns and is dissolved by the merciless plant. The dissolved juices help the plant grow in the nutrient-poor bog.

The seaside

WHERE THE LAND meets the sea – that is the seaside.

CLIFFS AND ROCKY SHORES

IF THE LAND is very high and rocky, there will be cliffs where it meets the sea. The beating of the waves against the cliffs can cause caves to form, or even sea arches and sea stacks. If the rocky land beside the sea is not high at all, then there will be a rocky shore. Here there will be large amounts of rocks and stones, many of them covered with seaweed, with lots of ROCK POOLS in between when the tide is out.

BEACHES, SALT MARSHES & MUDFLATS

IF THE LAND beside the sea is flat and sandy, that is where the beaches will form. The sea can move the sand up and down the shore, so the beach can look quite different after a big storm. If a flat piece of coast is protected by a ROCKY HEADLAND, this really slows down the sea currents. This means that mud instead of sand is brought in, and we have salt marshes and mudflats instead.

THE TIDE IS THE BOSS

THE SEA CONTROLS everything. The tide comes in and out every 13 hours, meaning it happens twice a day. This is because the moon exerts a force on the movement of the water. This is strongest when there is a full moon, every 28 days or so. At this time, the water covers the shoreline right up to the HIGH-TIDE MARK. Then it drops back again, taking over six hours to get to the LOW-TIDE MARK. Anything that lives on the shoreline has to be able to put up with being drenched with salty water twice a day.

Seaside plants

Seaside plants have to put up with salty sea water and salty spray. Most plants cannot do this, but those that grow well by the sea have specially adapted to be able to do so. Others just hang on, with their sides facing the sea damaged and prevented from growing by the salty winds.

Seaweeds

Seaweeds are well able to grow and reproduce in the sea. They don't have roots, but they have a grip at the base that holds on tightly to the rocks and stops the tides from washing them away. This is called a **HOLDFAST**. They grow in shallow water, on the rocks of the rocky shore and in rock pools. When they eventually die and let go, they are washed up on the sandy beaches, sometimes covering a big area if it has been stormy. There are three sorts – green, brown and red – and their Latin name, *algae*, is used when describing them. Seaweeds are the only plants that can grow on a rocky beach or on the lower parts of the sea cliffs.

Grasses and flowers

As well as salty spray, plants on the sandy shores have to put up with drought. The sand cannot hold water, so plants here have to grow really long roots to find the fresh water far below. The huge network of roots produced by **MARRAM GRASS** and **SEA COUCH GRASS** traps loads of loose sand and causes sand dunes to grow.

Behind these dunes, conditions are easier, and lots of yellow flowers appear in June. There are **SAND PANSIES**, **KIDNEY VETCHES**, **BIRD'S-FOOT TREFOILS** and **HAWKBITS**, as well as lots of different sorts of orchids – many of these are purple in colour. Blankets of fragrant purple thyme flowers cover the dune slopes all summer long.

Salt marshes are large flat muddy places that are completely covered by the tides at least once a month, so every plant there has to be able to survive this salty bath. Very few flowers can do this. Amazingly, many of those that do are coloured purple and flower in August.

The sheets of purple flowers here at this time include **SEA LAVENDER**, **SEA ASTER** and **SEA SPURREY**.

GLASSWORT, with its swollen stems full of fresh water, is able to grow right at the edge nearest the sea because it has its own supply of non-salty water.

Cliff flowers

The higher parts of the cliffs are affected by the salty sea spray, and very few flowering plants can put up with this. One that does is called **SEA PINK**. This grows anywhere there is even a small bit of soil, and it covers the clifftops in carpets of lovely pink flowers in May. A few others can deal with the salty conditions, such as **SEA CAMPION**, which has a white flower. **ROCK SAMPHIRE** has fat, swollen leaves in which it stores fresh water it gets from the rain, so it can put up with the salty spray.

Seaside invertebrates

Invertebrates are animals without backbones. They include insects (six legs), arachnids (eight legs, such as spiders), molluscs (no legs, such as slugs, snails, cockles and mussels), worms (no legs, such as ragworms and lugworms) and crustaceans (ten legs, such as shrimps and crabs). Vertebrates are animals *with* backbones.

Butterflies and moths

Sand dunes are covered in flowers in summer, so they are a good place for moths and butterflies to live. Bird's-foot trefoil is the food plant for the caterpillars of the **SIX-SPOT BURNET MOTH**. These yellow-and-black caterpillars turn into red-and-black day-flying moths, but because of the warning colours of both the caterpillar and moth, birds do not eat them.

The **SMALL BLUE BUTTERFLY** lives on sand dunes too – its caterpillars feed on kidney vetch.

SMALL COPPERS, **COMMON BLUES**, **GRAYLINGS**, **DARK-GREEN FRITILLARIES** and **PAINTED LADIES** are all butterflies that can be seen on nice sunny days.

Sand-dune snails

Sand dunes are great places for snails, as there is plenty of calcium in the grasses and leaves on which they feed to ensure that they have strong shells. Many of the sand-dune snails have lovely stripey shells. Thrushes hunt for these, but they need a stone on which to smash them open. So a stone or rock in the dunes is often surrounded by broken snail shells because it has been used as a thrush's anvil.

Mud-dwelling filter feeders

Mudflats are large flat muddy areas covered by the tide twice a day. No flowering plants can grow here, so you would think they are not good places to see wildlife. But they actually are! This is because under the mud lives a great collection of creatures. They get their food by filtering particles out of the sea water when the tide is covering the mud. And in turn, they are all food for huge numbers of wading birds.

The top few centimetres of mud is where the ***HYDROBIA*** snails and ***COROPHIUM*** shrimps live.

These creatures are as small as grains of rice, and there can be thousands of them in a few square centimetres. Deeper down in the mud is where the **COCKLES** and the **RAZOR SHELLS** hang out. Their bodies are protected by two tightly closed shells. When the tide is in, they open their shells and stick up long tube-like siphons to suck in their food. **RAGWORMS** and **LUGWORMS** live here too. They eat the mud and sand, digest any food in it, then excrete the cleaned sand through their bottoms up on to the surface, where it looks like strands of sandy spaghetti.

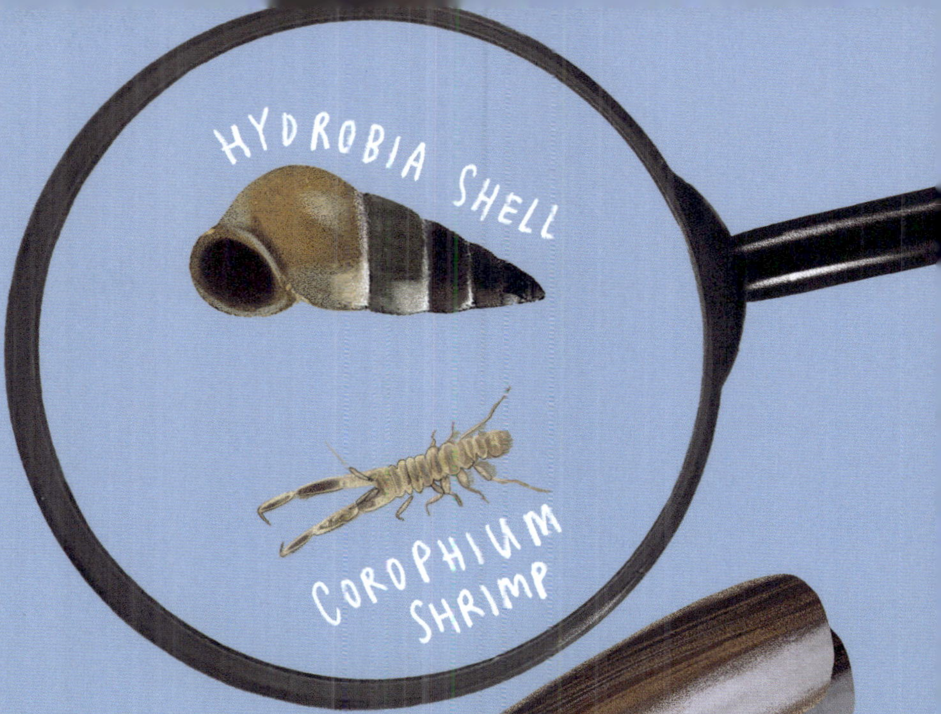

Sandy shores

There are creepy-crawlies on the sandy shores too, living along the high-tide mark. This is where you will find the largest woodlouse in Ireland, the **COMMON SEA SLATER**. It is more than twice the size of a garden woodlouse and can roll into a ball if it is disturbed.

If you turn over a heap of dead seaweed, you might also disturb hundreds of small creatures that go hopping off, frantically looking for shelter. These are **SANDHOPPERS**, and they are feeding on the dead seaweed – or they were, until they were interrupted.

Seabirds

Two seaside habitats are really important for birds. In summer, it's the sea cliffs, where seabirds come ashore to breed. In winter, it's the mudflats, where visiting waders with long bills come to feast on the creatures buried in the mud, and ducks and geese come to feed on algae and sea grass.

Sea cliffs

Seabirds can swim and they depend on fish for food, so they spend most of the year far out at sea. But laying eggs at sea is not a good idea, so they have to come ashore to breed and raise chicks. Sea cliffs with their rocky ledges are ideal for this, as seabirds like to nest together in great big colonies. There is safety in numbers.

Bird city

A busy cliff colony in early June will be home to thousands of seabirds. Up to 10 different species could be breeding there, but they are not just mixed up any old way. Each species has a particular part of the cliff where it nests in large groups. **FULMARS** arrive first and pick nest sites very close together at the top of the cliff – one pair on each. The **KITTIWAKES** – which are small white-and-grey gulls – are next and they huddle together in pairs as well. On the bare shelves underneath these are **RAZORBILLS** and **GUILLEMOTS**, which are black looking birds with white tummies.

These lay just one egg each on the cliff shelves and hundreds of them huddle very close together while they are incubating them for several weeks. The big black **SHAGS** have their nests below these, where the cliffs are a bit wider.

PUFFINS nest underground in burrows on the flat grassy places right on the cliff top. **GANNETS** – our largest sea birds – are so big and noisy and smelly they have to live by themselves on a different cliff face.

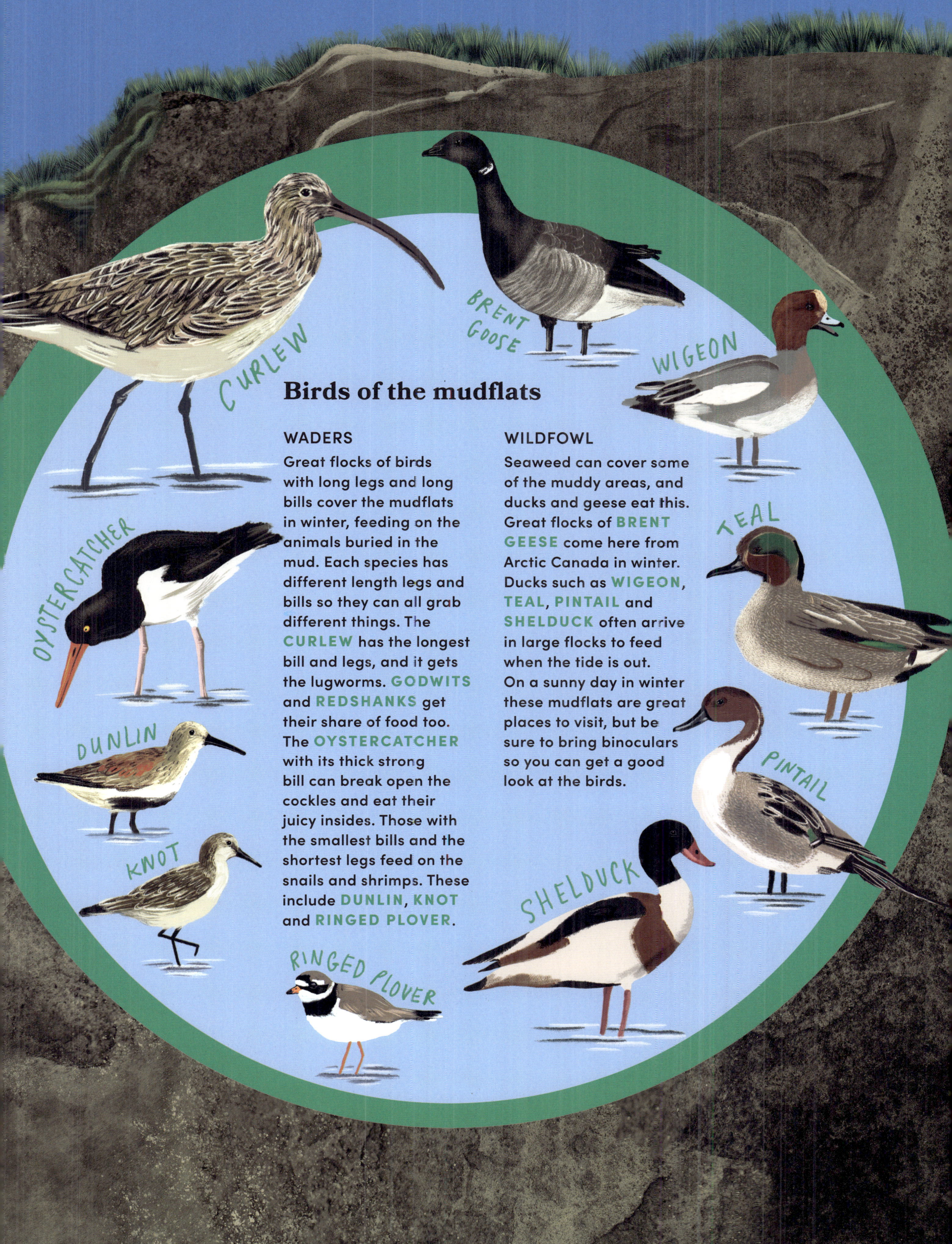

Birds of the mudflats

WADERS

Great flocks of birds with long legs and long bills cover the mudflats in winter, feeding on the animals buried in the mud. Each species has different length legs and bills so they can all grab different things. The **CURLEW** has the longest bill and legs, and it gets the lugworms. **GODWITS** and **REDSHANKS** get their share of food too. The **OYSTERCATCHER** with its thick strong bill can break open the cockles and eat their juicy insides. Those with the smallest bills and the shortest legs feed on the snails and shrimps. These include **DUNLIN**, **KNOT** and **RINGED PLOVER**.

WILDFOWL

Seaweed can cover some of the muddy areas, and ducks and geese eat this. Great flocks of **BRENT GEESE** come here from Arctic Canada in winter. Ducks such as **WIGEON**, **TEAL**, **PINTAIL** and **SHELDUCK** often arrive in large flocks to feed when the tide is out. On a sunny day in winter these mudflats are great places to visit, but be sure to bring binoculars so you can get a good look at the birds.

Seaside and sand-dune animals

Mammals that live in the seas around our coast are well-adapted to their environment. They have huge layers of fat called **BLUBBER**, which keeps them warm in the cold water. They can hold their breath for long periods of time, which means that they can hunt for fish underwater no bother.

Seals

Our two species of seal – the **GREY SEAL** and the **COMMON SEAL** – can often be seen from the shore, with their dog-like heads above the water. They are excellent swimmers, as their back legs are actually flippers that they can rotate to change direction and help them steer as they swim.

These back legs are no good at all for walking, so when the seals come on land to have their pups and visit them for feeding, they have to drag themselves forward using only their forelimbs. They look very awkward indeed. The grey seal is bigger and more common than the common seal – who thinks up these names?

Whales and dolphins

Whales and dolphins belong to a group of mammals called **CETACEANS**. While they have lungs and must come up to the surface for a breath every so often, they never come ashore. They feed and sleep in the sea. They mate, and their calves are born, underwater.

The sea water supports their great weight and, in fact, if they are washed on to the beach alive, their lungs will eventually be crushed by the weight of their ribs and chest because there is no water to keep them buoyant.

Rabbits

Sand dunes suit rabbits very well. They can easily make burrows in the sandy soil, and there is lots of grass for them to eat in the fixed dunes further back from the sea. Grass is hard to digest, so rabbits eat their own poo and have a second go at digesting all the nutrients in each meal of grass.

There are always lots of rabbits because each female can have about five babies per litter and up to seven litters in a year – 35 babies in all. Just imagine!

BOTTLE-NOSED DOLPHINS, **COMMON DOLPHINS** and **PORPOISES** can be seen leaping through the waves close to shore, as these are our most common species. Bigger whales keep offshore, but they blow spumes of air through blowholes and some, such as **HUMPBACK WHALES**, come right up to the surface and wave their tails in the air as they dive.

Lizards

We have only one species of lizard in Ireland – the **VIVIPAROUS LIZARD** – and they love sand dunes. Because they are not mammals but reptiles, which are cold-blooded creatures, they need to bask in the sun to get warm enough to be able to move, so they like to lie around sunbathing. They eat insects and slugs and snails. Kestrels, crows and stoats are their enemies. If caught by the tail, they can break it off to escape and grow another one later.

Rock pools

Rock pools are wonderful places to find seashore animals. A small net is useful, as you can lift things out and put them back safely afterwards. Low tide is the best time to explore, when the rocky shore is dry enough to walk on, but the pools are still full of sea water and life.

Fish

When you look into a fairly big rock pool you may see a whole shoal of tiny fish swimming fast together. If you try to catch them, they will all dash away under the nearby seaweed. But if you try harder, you might catch some in the net.

These are usually shoals of the **TWO-SPOTTED GOBY** (although only the female has two spots – the males have four). An eel-like fish with a sad-looking face that can be found by itself down at the bottom is the **BUTTERFISH**.

Shellfish

Rock pools are great places for crabs. **SHORE CRABS** are quite small and can be any colour from yellow, to red, to green, to black. They hide in the seaweed and eat anything they can get their claws on – no matter whether it is dead or alive.

HERMIT CRABS live in empty shells of sea whelks. If you lift one out of the rock pool you can see its eyes looking out at you. The see-through creature swimming backwards at great speed is the **COMMON PRAWN**. It has pincers on the front two pairs of its legs.

Pretending to be flowers

If you look at the sides of the rock pools you will probably notice blobs of red jelly stuck to the walls. But look carefully at the same blobs under the water and they will have opened up and look like flowers. These are not flowers, however – they are hungry animals called **SEA ANEMONES**.

Their 'petals' are a row of tentacles that surround their mouth. They make a current by waving their tentacles and they suck in any small creatures caught in the current. If you put your finger into the anemone, between the tentacles, you will feel the current too.

Protected by shells

Hard shells give great protection to the animals inside. **BARNACLES** are small, pointy and white and are permanently stuck to the rocks. They have diamond-shaped trapdoors at the tops of their shells. When the tide is in, they open these doors and use their feet to kick food into their mouths.

LIMPETS, which are much bigger, are also stuck on the rocks around the rock pool. But when the tide comes in, they let go and walk around, scraping tiny algae off the rocks.

PERIWINKLES come in different colours and sizes – from yellow to black. These are sea snails that live on the bottoms of the rock pools. If there is a lot of seaweed in a rock pool, you might even find

MUSSELS, which have two shells clamped tightly shut. They hold on to the seaweed with special byssus threads that poke out from between their two shells.

A seaside walk

Anyone can walk along the coast because all the land on the sea-side of the high-tide mark is public property. Sandy beaches are particularly enjoyable places to walk, and they are always changing. You never know what interesting thing may have been brought in by the tide since you were there last.

Seashells on the sandy seashore

You can always find empty shells washed up by the sea. Empty cockle, mussel, razor and clam shells will be quite abundant. Look out for empty **WHELK SHELLS**, and if you find one hold it up to your ear, like a phone, and you will hear the sea through it. **PELICAN'S FOOT SHELLS** and **OYSTER SHELLS** occur on some beaches too. If you are really lucky you might come across a whole sea **URCHIN SHELL** or even a **SEA POTATO**.

Flotsam and jetsam

These are other things washed in by the tide that are no longer alive. You should find common jellyfish, which are the size of fried eggs, and mermaid's purses, which are the empty egg cases of rays and skates. Pieces of timber that have been in sea for ages may be covered with black and white **GOOSE BARNACLES**.

People thought long ago that these grew into barnacle geese because they are the same colour. The rarest thing of all is a grey substance called **AMBERGRIS**. This was formed in the guts of sick whales but is very valuable to those who make perfume. They would pay loads of money for even a small bit. It very occasionally gets washed up on our beaches.

Seashore birds

As you walk along a sandy beach you may see groups of little brown and white birds running in and out of the waves as they break on the shore. These are **SANDERLINGS** looking for food. A gravelly beach may well have **TERNS** nesting – such as the Arctic and the common terns. Look, but don't disturb them by walking through the colony. Gulls are always waiting to grab eggs and nestlings if the terns are not there to protect them.

Beached cetaceans

Sometimes – sadly – a dolphin or whale gets washed up on the beach and is stranded there when the tide goes out. It may be a sick animal that came in too close to the shore because it could no longer navigate its way. It may even still be alive, and if this is the case, it can sometimes be refloated on the next tide. Or it may have died at sea and been washed up. The biggest animal ever to live on earth is the **BLUE WHALE**. In 1891, one was washed up in Wexford, and its skeleton is now in the British Museum in London.

Gardens and parks

ARE PARKS AND GARDENS REALLY WILDLIFE HABITATS?

PARKS AND GARDENS used to be entirely man-made and strictly managed. The area to be the garden or park was marked out and all the wild plants removed. They were called rude names, such as weeds, and dug up and destroyed. Then they were replaced with plants and trees that people thought were nicer to look at. Evergreen hedges of box or laurel were planted around the boundaries of the gardens. Exotic trees from North America or China were planted in the parks. And flowering plants were bought in garden centres and laid out in beds. Any insects that dared eat any of these were called pests and killed with sprays and poisons.

SO ARE THEY GOOD FOR WILDLIFE?

A GARDEN FULL of red roses, tulips and peonies all sprayed to kill greenfly are not good at all. Bees cannot see the colour red – to them it looks black – so they never visit these red garden flowers. But people are more aware now of how important it is to have spaces for wildlife, so things are changing. New garden hedges are likely to contain native bushes such as hawthorn or holly. Garden vegetables are less often sprayed, as people don't want to eat food that has been treated with poison. We know how essential it is to have lots of bees so we will have apples on our apple trees. Grassy areas in parks are mowed less often so that wildflowers can grow.

In fact, gardens and parks can be excellent places for wildlife. We can see wild things from our kitchen windows or on walks around the local park. All you have to do is keep your eyes open and stay quiet so as not to scare them away.

Garden plants

Biodiversity begins with plants, and this is just as true in a garden or park as in a wild wood or on a bog. Many of the flowers and vegetables that have been planted in a garden for the benefit and enjoyment of the gardener and their family are actually excellent for wildlife too.

Growing garden crops

Plants must be **POLLINATED** in order to set seeds. The pollen – which is the male part – must land on the female part of the flower and fertilise it. Seeds are formed after this. While the wind blows the pollen from grass plants and tree catkins, insects do the job on flowering plants. They don't know they are doing this, of course.

Flowers repay the insects with nice food – nectar and, in the case of bees, pollen too. The nectar is hidden deep in the flower. The insect must stick its head right into the flower to get it, and its head becomes covered with pollen as it does. The insect then flies off to another flower to get more, bringing the pollen along on its head.

The insect has no notion that it is pollinating flowers – it thinks it is just getting tasty drinks.

Gardeners that want peas, or strawberries, or apples, or tomatoes won't have any if there are no pollinating insects.

Nor will they have any seeds to plant carrots or lettuce or onions if no pollination occurs. Clever gardeners are happy to have lots of pollinating insects in the garden to have good crops. They would never dream of killing them by spraying.

Wildflowers in gardens and parks

Wildflowers are the native plants that grew before the gardens and parks were set up. Lots of native wildlife depend on them. So allowing wildflowers to grow in parts of the garden is really good. The seeds are still in the soil and will grow if left alone.

Not cutting the lawn during May allows lots of wildflowers to grow. Leaving a bare patch of soil will do the trick too. Wildflowers such as **SELF-HEAL**, **CLOVERS**, **SPEEDWELL** and even **PYRAMID ORCHIDS** are only dying to get the chance to grow.

Garden insects and other creepy-crawlies

Gardens and parks full of flowers are great places for insects from one end of the year to the other.

Bees

The large **BUMBLEBEE** queens wake up from hibernation in March, and they are absolutely starving ... They need a feed of nectar quick so they can start building a nest, laying eggs and getting the bee colony going. They absolutely hate when all they can find are daffodils. These garden flowers have no nectar or pollen.

Daffodils don't depend on bees to pollinate them to make seeds. The bumblebee loves to find dandelions or grape hyacinths, crocuses or even snowdrops. All summer long, bumble bees, solitary bees and honeybees will visit the garden to feast on nectar and pollen.

The food chain on roses

Roses are no good for nectar – they have so many petals that insects can't get in – but **APHIDS** love to suck the juices from their leaves. Aphids, in turn, are eaten by **LADYBIRDS**, such as the seven-spot, the two-spot, the ten-spot and, increasingly, the invasive harlequin, which can eat our native ladybirds as well as aphids.

LACEWINGS gobble up aphids too, as do small birds such as blue tits.

ANTS hang around the rose bushes with the aphids as well – not that they actually eat them. Instead, they feed on the sticky juices that come out at the backs of aphids after they have been eating the plant juices. The ants will actually protect the aphids so that their predators can't eat them.

Spiders

Hunting spiders run over the ground seeking anything they can grab to eat. The **WOODLOUSE SPIDER** feeds entirely on woodlice, which it can spear with its huge fangs. The **WOLF SPIDER** lives up to its name and can run very fast to chase small insects. The **NURSERY WEB SPIDER** is a night-time hunter. The female carries her egg sac under her body as she rushes around hunting and builds a nursery web for it when it is about to hatch out.

Centipedes

Centipedes are vicious hunters too. One species – called **LITHOBIUS** – is able to catch worms, slugs and even other centipede species. It kills them with the poison claws behind its head. Centipedes never have a hundred legs – always fewer. Scientists long ago mustn't have been able to count.

Cuckoo spits

In June, lots of plants like lavender, robin-run-the-hedge and brambles are suddenly covered in white globs called cuckoo spits. Because the cuckoo was around in June, people once thought that these spits were left behind by the cuckoo as he cleared his throat between calls of 'Cuck-oo'.

But if you lift one on to your finger, you will see that it contains a tiny green insect with lovely brown eyes. This is a **FROGHOPPER**. The egg was laid on the leaf by the mammy froghopper and covered with this lovely sugary froth. This is food for the insect when it hatches out. It must be like living in the middle of a 99 ice cream!

Garden birds in spring and summer

Gardens are great places to see birds at any time of year. BirdWatch Ireland, which organises a garden-bird survey every year, has recorded 110 different species that have been seen in Irish gardens. Number one on the list is the robin.

Robins

The male robin starts to sing in spring, and this means he has decided that your garden is now his territory. The singing keeps away all the other males and attracts a female. Although, as females look exactly like males, he may be confused at first and try to chase her away.

But he soon gets sense, and the pair build a nest in a sheltered place in the garden. They can rear two clutches of five each over the summer. Young fledged robins don't have red feathers but have brown speckled ones instead. This is so Daddy doesn't see them as a threat to his territory.

Tits

Three of our four tit species – the **GREAT TIT**, the **BLUE TIT** and the **COAL TIT** – live and nest in gardens. They nest in holes in trees in woodlands, but by putting up a nest box with the right-sized hole in your garden, you can get tits to nest there. They love collecting caterpillars and greenflies to feed their nestlings – a busy job, as there can be up to eight in a nest box.

Magpies

Magpies build their nests early in the year, and like to nest at the top of tall trees. They will steal eggs, and little nestlings, from smaller birds' nests to feed their own young.

So good cover for your garden birds' nests – such as thick ivy on the wall or a dense spiny hedge or, indeed, nest boxes – means that nesting birds are safe in your garden and the magpies can't get them.

Dawn chorus

Did you know only male birds sing? As the days get longer in April and May the male birds are stimulated by this extra light to start breeding behaviours such as singing. All the different species sing together when it begins to get bright – sometimes as early as four o'clock in the morning. This half hour of birdsong, in an otherwise quiet early morning, is called the dawn chorus. By leaving the bedroom window open, you can hear it all without even getting out of bed.

In a good garden, you might hear **WOOD PIGEONS**, **WRENS**, **THRUSHES**, **BLACKBIRDS** and **FINCHES**, as well as robins and the tit species. When it gets really bright, they all fly off to look for food and the chorus is over.

Garden birds in winter

Many of our songbirds depend on our gardens to provide them with food to survive the winter. The food can be natural, such as berries and crab apples, or it can come from our well-filled bird feeders.

Blackbirds

These birds depend on berries to get them through the winter. They love the red berries of the mountain ash and the yellow berries that grow on pyracantha bushes. They love apples too and will happily feed on the crab apples that are around at this time. They cannot manage hanging bird feeders – they feed on the ground, poking their bills into grassy areas for worms or eating apples cut in half that we leave on the ground for them.

Thrushes

Several species of thrush visit our winter gardens. The **SONG THRUSH** is the only one that can eat a snail. It holds it in its bill, bashes it off a stone to break the shell and then gobbles up the contents. The **MISTLE THRUSH** loves holly berries and you often see one defending a nice berried tree. It will sit there and chase off any other bird that tries to snatch one.

Migratory winter thrushes can be found eating the berries too. **FIELDFARES** and **REDWINGS** spend the summers in northern Europe but come here in winter because it is warmer. Redwings can occur in large flocks, which you might see in the park. The groups of fieldfares are smaller, and they sometimes join up with the redwings. Check out any flocks with your binoculars.

Sparrowhawks

This is a bird of prey that feeds on small birds. It can come swooping in when the birds are busy on the feeders and grab a finch or a tit before they realise it's even there. Female sparrowhawks are larger than the males, and they can kill and eat pigeons and magpies if they get a chance to grab one. She grabs it in her sharp claws and quickly kills it. Then she plucks off the feathers and eats the tasty bits first, like the brain and the liver. She often flies off with it then, leaving a mass of feathers behind.

Waxwings

The most exotic bird you can see in the garden or park in winter is the waxwing. They are pink-brown in colour with a prominent crest on the head. They have yellow and white tail and wing feathers, with bright red dots on the wings. They sound like musical bells as they fly about in small flocks. They come from Scandinavia but not every year. They can strip a bush of berries in a few hours, as they can eat twice their body weight per day.

Garden animals

We may think our gardens belong to us, but several mammals consider that they belong to them as well. They are part of the natural food chain of the garden, feeding on the smaller creatures that live on garden plants. They also will eat any food provided by the human owners.

Foxes

Foxes are the animals most commonly seen in gardens – especially in urban areas. They are not afraid of people and come in looking for food – either stealing it out of the dog's dish or digging holes for worms. Their home is called a **DEN**, and they often build these in bushy areas in parks. These are very messy because foxes are untidy and leave bits of food scattered at the entrances.

Foxes mark their territory with very smelly urine to warn other males to keep away. Fox cubs, born in spring, often play in sunny back gardens while their parents look for food for them.

The female is called a **VIXEN**, and she communicates with the male in winter by making blood-curdling screams that sound like someone is being murdered. These screams might be the origin of the Irish banshee – a mythical female spirit whose wailing warns of a death.

Hedgehogs

Everyone loves to have hedgehogs in the garden. Gardeners like them because they eat slugs and snails that would otherwise be eating their garden plants. The rest of us like them too and are always amazed to see them tightly rolled into a ball. They do this when they feel they are in danger. The only animal that can unroll and eat them is the badger.

Hedgehogs **HIBERNATE** for the winter. When they feel the days getting colder, they find a nice sheltered place and curl up there, covered with fallen leaves, and go into a deep sleep. The warmer weather wakes them up. It might be April before this happens – there's no point in waking up in February when there will be no snails or slugs for them to eat. Baby hedgehogs are born in June.

Their prickles are very soft at first. It takes a few weeks for them to stiffen up so that they will protect the young hedgehog.

Hedgehogs were brought to Ireland by the Vikings, who used to eat them as snacks – having killed and roasted them first. Irish people called these new animals *an gráinneog* – 'the ugly one' – which is their Irish name to this day.

Rats

You only get rats where there are people. They live near to us and eat any rubbish we leave around. The rat species we have now is the **BROWN RAT**. This came to Ireland at the beginning of the 1700s on ships. Before that we had the **BLACK RAT**, which carried the Black Death in the 1300s.

Nearly half the population at the time died from this disease, but now the black rat can only be found on Lambay Island. Brown rats carry other diseases, so nobody wants them around. But they are very clever, and we have lots of them, although they mostly only come out at night. Foxes love to catch and eat them.

The garden at night

Garden birds, all of which feed during the day, have to go to bed when it gets dark. But night is a very safe time for lots of other animals, so it is worth inspecting the garden with a torch after the sun goes down. There is a whole different world out there then.

Snails and slugs

It is much cooler and damper at night, so animals with soft slimy skins are in no danger of drying out. Look at the leaves of plants such as hosta or lettuce or basil and you will surely find lots of these molluscs feasting on them.

GARDEN SNAILS have large brown shells on their backs. Like all molluscs, they have four horns on their heads – the eyes on the top two and their nose on the bottom two. All their teeth are on their tongue, and they scrape them over and back to eat away at the leaves.

Slugs are just as hungry. There may be several species of slugs around, such as the **LARGE BLACK SLUG**, the **GREAT GREY SLUG** and the really hungry **SMALLER GARDEN SLUG** that eats lots of young, tender plants under cover of darkness.

Woodlice

These have no protective coating to stop them drying out, so they prefer night-time as well. They hide under flowerpots and logs during the day. They feed on dead and decaying vegetation and are really useful, as they help to recycle nutrients back into the soil.

There can be hundreds of the **COMMON WOODLOUSE** wandering about at night. The **PILL WOODLOUSE** is common too. If you pick it up, it will roll into a ball – the same defence strategy the hedgehog uses – which confuses the would-be predator.

Owls

Owls are the only birds that hunt at night. Large wooded parks can be home to **LONG-EARED OWLS** who fly very silently, listening out for the tiny sounds made by scurrying mice and rats on the ground. They pounce from the air and never miss. They detect these sounds through large discs on their face. The 'long ears' are not ears at all – just long feathers on top of their heads.

Bats

Bats and birds both catch flying insects. Birds are around during the day, and bats take over at night. Street lights can attract moths and, in turn, these are caught by **LONG-EARED BATS** who are hunting in the area. These bats also fly along wooded sections of parks, feeding on any flying insects they detect with their echolocation signals.

Beetles

Big black beetles come out at night too. The **DEVIL'S COACH-HORSE** is very impressive. It is up to 30 millimetres long, and when it thinks it might be in danger it doesn't run away. It turns to face the enemy, raises its tail to look like a scorpion, emits an awful smell and opens its ferocious-looking jaws. That puts off most predators. In fact, it can even bite humans if we pick one up. Other black beetles are running around too, with violet reflecting parts to their bodies.

Turn your garden into a wildlife haven

The trick to making your garden a wildlife haven is to have several different habitats to increase biodiversity. As well as providing the right plants and trees, you can provide food for visitors. Hedgehogs like meaty dog food. Bird feeders and bird tables are great for birds in winter.

A mini-woodland

Just one tree, some shrubs and a stretch of hedge will suit much woodland wildlife. If your tree is a native deciduous one with flowers in spring, lots of young, tender leaves in summer and fruits or nuts in autumn then wildlife will love it.

The hedge will be cover for nesting birds, hedgehogs and shrews. Shrubs with lots of flowers and berries, not necessarily wild ones, are all great too. Birch, rowan, hazel, holly, hawthorn, pyracantha and cotoneaster are all excellent.

Make a pond

Even a very small patch of shallow open water attracts lots of different living things. A hole lined with a special pond liner and planted with a few water plants is all you need. Much of the wildlife will arrive by itself by flying in. If you fill the pond with water from another pond, that will bring other wildlife. Never use tap water. Don't put in fish either, or they will eat everything else.

FROGS do very well in garden ponds. Mating adults may be attracted there in spring and will spawn after mating. Soon the tadpoles will hatch out of the jelly and swim around feeding on other insect life – or you can feed them with fish food.

They will grow quickly, developing their hind legs first, followed by the front pair. This can take 12 weeks, if all goes well, and then the new frogs will hop out and leave the pond. They will return to spawn the following spring.

Create breeding habitats

Several species of birds, such as tits, sparrows, robins and swifts, will breed in **NEST BOXES**. Don't put up a lot of boxes unless your garden is very large. Gardens are usually the right size to be breeding territory for only one pair of a species, such as blue tits or great tits.

Grow ivy or other climbing creepers on a wall – this can be enough shelter for breeding blackbirds or dunnocks. A log pile in a corner will be home for lots of creepy-crawlies, such as woodlice and millipedes. **BUG HOTELS** are great too.

Make a grassland

If you have a lawn, you can let some of it become an area of natural grassland by not mowing for six weeks at a time.

You can even make a **WILDFLOWER MEADOW** by planting native Irish wildflower seeds on a patch of bare soil and keeping the grass away.

Grasslands

GRASSLAND IS THE name given to the fields full of grass that cover 60 per cent of Ireland. When these fields are surrounded by hedgerows, it is called **BOCAGE LANDSCAPE**. This only occurs in Ireland, Britain and northern France, nowhere else in Europe. Grassland must be grazed by animals or cut by machinery for hay and silage – otherwise the trees and bushes would soon spread in from the surrounding hedges, and the whole place would be woodland again.

INTENSIVE GRASSLANDS

THESE FIELDS JUST have one or two species of grass and no other plants at all. They are given nitrogen fertiliser, and are often sprayed with slurry, to make the grass grow. The grass is cut right down for silage twice a year, so these fields contain very little wildlife and are not good for biodiversity.

MEADOWS

THESE ARE GRASSY fields grazed by animals. They contain more plant species, and the animal droppings are food for beetles and other creatures in the soil. They have a bit more wildlife but are still poor enough.

SEMI-NATURAL GRASSLANDS

THESE ARE PERMANENT pastures that are lightly grazed and contain lots of wildflowers and insects. The hedges that surround them contain native tree species and act as important wildlife corridors.

WET GRASSLANDS

IN FIELDS WHERE the drainage is poor, lots of rushes can grow. The plants will be different here, and there will be many varieties too. Farm animals can only graze when it is dry enough, so there is less pressure on the plants. This means they can live long enough to have flowers and seeds.

Grassland plants

Wide open spaces with no shading from trees are great for small flowering plants. If the soil is rich, lots of grass will grow. If the soil is too poor for these strongly growing grasses, other plants with nice-coloured flowers get a chance, as well as some more delicate grasses.

Grass

There has to be grass, of course, or it wouldn't be a grassland. There are over 200 species of grass in Ireland, and some are much better for biodiversity than others. **PERENNIAL RYE GRASS** is great for feeding cattle but little else, while **YORKSHIRE FOG** and **COCK'S-FOOT** are food plants for lots of caterpillars. Grasses have long thin green leaves, and their flowers grow on separate stalks. These flowers are small, have no nectar or petals but are full of pollen. The wind blows the pollen from one plant to the next, so there is no need for insects to do it. This is why on a warm summer's day the air is full of grass pollen that causes allergic reactions in people who suffer from hay fever.

Flowers

Some flowers that grow in grasslands work very hard to improve its fertility for other plants. These are called **LEGUMES**. They take nitrogen directly from the air, which hardly any other plants can do, and fix it in the soil. Then plants that might need more nitrogen can take it up and grow better.

CLOVERS – red, white and yellow ones – as well as **VETCHES** and **MEADOW VETCHLINGS** are all able to capture nitrogen for other plants. Then there are other grassland flowers that do as little work as possible and live off their neighbours.

YELLOW RATTLE is a semi-parasite, which means its roots steal nutrients from the roots of grasses. **BROOMRAPE** has no green leaves at all – it gets all its nutrients from nearby clovers.

Colourful wildflowers

A wildflower meadow in the month of May will have flowers of every colour. Yellow for buttercups, dandelions, rapeseed, lady's bedstraw, ragwort and cowslips. Purple for speedwell, milkwort, self-heal, scabious, thistles and the lovely, pale lady's smock. White flowers include daisies, white clover, stitchwort, mouse-ear and chickweed. Because bees cannot see red, there are hardly any red wildflowers. Poppies are actually more orange than red and have ultra-violet markings, which we cannot see but bees can, so they are an exception to the no-red-wildflowers rule. There are no black, brown or green flowers either.

Mushrooms

Semi-natural grasslands are great places for mushrooms in summer and autumn. These mushrooms live on the roots of grasses that have died, as they cannot make their own food. Some are white with pink gills underneath.

Others are big white balls that appear overnight near the hedges and were called sheep's farts when I was growing up. There are red ones too and purple ones, but never eat a mushroom until someone who knows tells you if it's safe to do so (and eats one first).

Grassland wildlife

Wildflower meadows in full bloom are full of biodiversity. There will be lots of flying insects as well as ground-living ones. Birds that nest on the ground will be there too, their nests safely hidden in the long vegetation. Grass-eating animals love these places as well.

Butterflies

On a warm sunny day in June, a grassland full of wildflowers will be full of butterflies. They are looking for the food plants that their caterpillars prefer in order to lay eggs on them after mating. **SPECKLED WOODS** lay on grasses such as Yorkshire fog and cock's-foot, as does the **WALL BROWN**.

The caterpillars of **MEADOW BROWNS** and **RINGLETS** love these grasses too. Really flower-rich meadows will have white butterflies such as **ORANGE TIP** and **GREEN-VEINED WHITE**, as well as **RED ADMIRALS** and **PAINTED LADIES**. And of course, all these butterflies visit the flowers for nectar for themselves during this busy mating and egg-laying time.

Bees

Bumblebees love nice grasslands. The queen will have a nest in the soil, and the workers will gather nectar and pollen from the meadow flowers to bring back there. Solitary **MINING BEES** each have an individual nest. Many make these by tunnelling in the sandy soil, and then they lay an egg or two inside.

After that, they stuff the nest tunnel with pollen for food and seal it up. The eggs inside are left to hatch out, the food is there for the young, and they get on with things themselves with no parental help at all.

Birds

Birds that nest in grasslands do so on the ground in the cover of the long grass. **CORNCRAKES** and **CURLEWS** do this, and they used to be very common indeed in the rich grassland areas.

But a field that is cut to the ground for silage in May or heavily grazed by hungry cattle provides no cover for ground-nesting birds. We have hardly any curlews or corncrakes left now.

SKYLARKS nest on the ground in poorer grasslands, and the males fill the air with their song as they defend their territory from above.

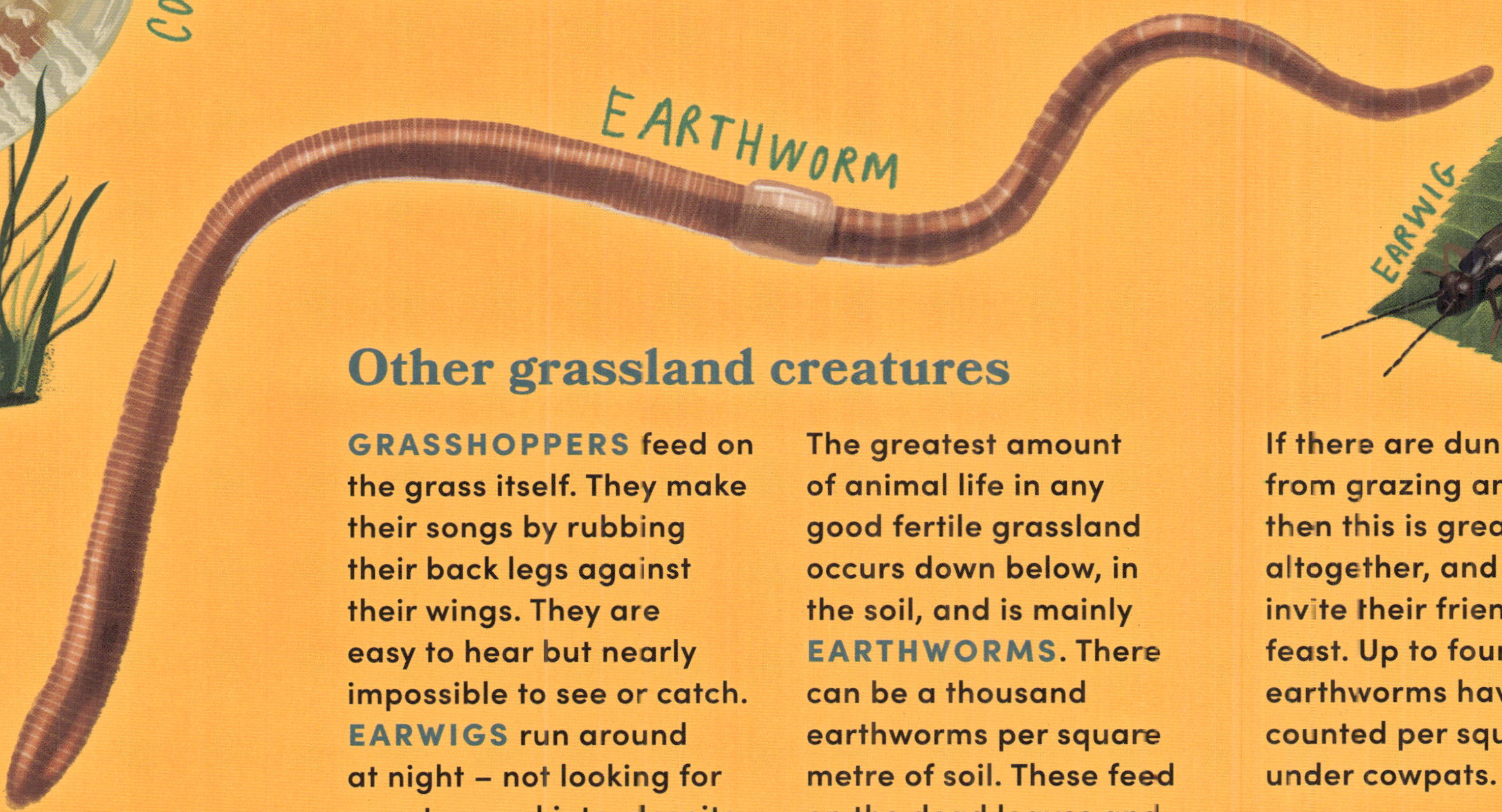

Other grassland creatures

GRASSHOPPERS feed on the grass itself. They make their songs by rubbing their back legs against their wings. They are easy to hear but nearly impossible to see or catch. **EARWIGS** run around at night – not looking for ears to crawl into, despite their name, but looking for insect larvae and flower petals, which they gobble up with delight.

The greatest amount of animal life in any good fertile grassland occurs down below, in the soil, and is mainly **EARTHWORMS**. There can be a thousand earthworms per square metre of soil. These feed on the dead leaves and roots of the grassland plants by dragging them down into the earth and eating them along with the soil itself.

If there are dung pats from grazing animals, then this is great altogether, and the worms invite their friends to the feast. Up to four thousand earthworms have been counted per square metre under cowpats.

Hedges

In 1751, landowners were required by law to put hedges or enclosures around their land. New crops and animals had been developed that needed sheltered fields. Lines of hawthorn or furze were planted to divide open land into fields. Soon other trees, such as hazel, holly, willow, elder and ash, grew up among them.

Biodiversity

These hedges greatly improved the wildlife biodiversity of the fields they surrounded. They formed **WILDLIFE CORRIDORS** for the mammals that live there. This means that badgers and bank voles and shrews and mice can move long distances without crossing fields and being exposed to enemies. The banks of earth that the hedges were planted on can be easily tunnelled for burrows for the rabbits that eat the grass in the field. **STOATS** live in hedges and can kill rabbits that are much bigger than them by jumping at their throats and biting them. **NATTERER'S BATS** hunt along the hedges at night, catching their fill of the moths that also live there.

Badgers

Badgers live in holes called **SETTS**, which they dig into the banks of hedgerows. A family of badgers may live there for centuries, as the sett is passed down from mother to daughter. They do not rush out of their setts and grab the leg of a passer-by with their strong jaws, never letting go until the bone breaks. People used to believe this and would carry a stick when walking along a hedge. They kept it ready so they could snap it in half and fool any attacking badger into thinking that the noise of the breaking stick was the noise of their leg-bone breaking. Such nonsense!

Birds

The trees and bushes of the hedges provide cover for nesting birds, such as yellowhammers, dunnocks, linnets and wrens. **YELLOWHAMMERS** and **LINNETS** were very common when there were more fields growing crops of barley and oats and the grass grew long enough to have grass seeds from their flowers.

In autumn, great crops of berries, such as rosehips, haws, blackberries, elderberries and holly berries, attract **MISTLE THRUSHES**. They also are a feast for the **REDWINGS** and the **FIELDFARES,** which are migratory types of thrush that visit us from mid Europe during the colder months.

Dead birds and burying beetles

When birds get old and weak, they often go into hedges to shelter and die. Then a pair of red-and-black beetles, which are called burying beetles, come to bury the body. They work very hard digging a hole under the dead bird until it sinks below the ground. The female then lays eggs in the corpse, they cover it with soil and go away. The young beetles emerge eventually and feast on the carcass.

THE GREAT IRISH NATURE CHALLENGE

Robin's pincushion

This is a weird-looking red growth that occurs on the leaves of a wild rose plant. It is caused by a tiny wasp called *Diplolepis rosae*. This solitary wasp lays an egg in the leaf in summer. When the larva hatches out from the egg it produces chemicals that make the leaf grow into this shape and colour.

This is the robin's pincushion GALL, and the young wasp lives inside it until the next spring. It then emerges as a flying adult, mates and goes off to lay eggs on other wild rose leaves. It doesn't actually damage the rose plant.

Orchard ermine moth

A hawthorn hedge covered in webs looks particularly spooky, especially when you see it in June rather than at Hallowe'en. Spiders are not to blame, however. Examine it closely and you will see it is full of thousands of caterpillars that have eaten all the leaves off the hawthorn and covered themselves in this sticky web. If you touch one on the head it will wriggle backwards. By July, these will all have become lovely white adult orchard ermine moths. They will leave, mate, lay eggs and die, and the young will overwinter as tiny non-feeding caterpillars. The hedge recovers completely once the adult moths leave.

Ash trees

As many of our hedges are around fields with good fertile soil, they contain lots of ash trees that grew there naturally from self-sown seeds (called keys). So, no problem finding ash trees. But look carefully. Is the tree healthy? Many, many of our ash trees are now suffering from a disease called **ASH DIEBACK**, which kills them and can't be cured.

This disease is caused by a fungus-like organism that arrived here from mainland Europe by 2012. Its spores blow from one tree to the next, spreading the disease. The only hope is that we may have some resistant trees, so see if you can find a healthy ash tree.

Remember, these plants and animals are much rarer than others – how many can you find?

Barn owl

Barn owls are large, white owls that hunt along hedges at night. They build nests inside old farm buildings or ruined churches in old graveyards. Their call is a long, drawn-out eerie shriek that terrifies people who don't know what it is. They catch lots of rats during their night-time hunting to feed their four young. Farmers put up huge nest boxes to attract them, which look like dog kennels up in trees. They are very rare so it's a great thrill to see one.

Rivers and lakes

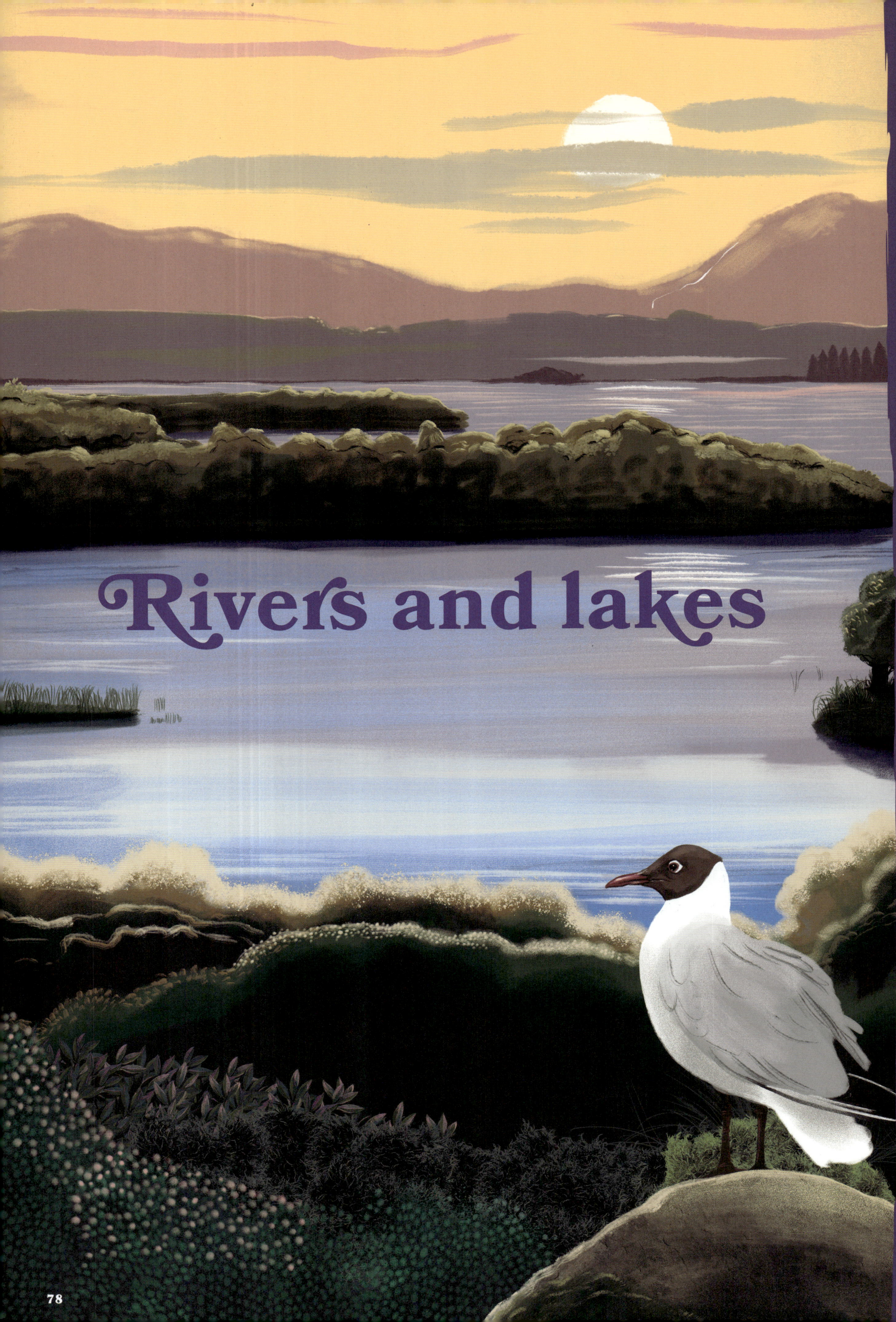

ALL OUR RIVERS and lakes come from the rain. It rains a lot in Ireland, and there is more rain on mountains than in flat areas. The rainwater runs off the sides of the mountains in small streams that meet up and grow into rivers. They flow downhill and are joined along the way by streams of water coming off the land they pass through. Where the land is steep, rivers flow fast. Where it is less steep, the rivers broaden out and flow slowly. If the land is nearly flat, the water practically stops, widens out and becomes a lake for a while. But eventually every river reaches the lowest point, which is the coast, and then it flows into the sea.

IS IRELAND'S WATER CLEAN?

THAT ALL DEPENDS on what is happening here. Freshwater habitats should have lots of biodiversity and be full of wildlife. There should be plenty of oxygen in the water for the animals living there to use. Our rivers and lakes are tested all the time (by the Environmental Protection Agency – the EPA) to see what state they are in. Ireland can be fined by the EU if there is so much pollution in the water that it is hard for wildlife to live there. In fact, in 2012 Ireland was fined €2.7 million because faulty septic tanks were polluting the water.

WHAT IS CLEAN WATER?

EVERYONE WOULD SAY that clean water is water that fish, insects and plants can live in. Yet if you turned on the tap for a glass of water and an insect or a tiny fish went into the glass with the water, would you drink it? Would you prefer completely clear water without even invisible bacteria in it? That is what comes out of our taps. The water has been treated with small doses of chlorine that don't harm us but kill all the bacteria present, so nothing can live in it. So what is clean water? The answer is it all depends on what the water is for.

Trees and plants

Most of our freshwater plants grow in and around lakes. This is because the water in rivers moves too fast for plants to get a grip.

Deep water

WATER LILIES grow in deep lake water, with their roots in the soil at the bottom. Their flat leaves on the surface are connected by a very long stem.

In summer, the flowers come – white in acid-water lakes in mountainous areas or yellow in lowland calcium-rich waters.

Willow

Willow can grow very well here too. We have four native types of willow, and each one has lots of biodiversity that depends on it for food. Willow **CATKINS** in spring provide pollen and nectar for early bees. Caterpillars, such as those of the **SALLOW KITTEN MOTH** and the **LUNAR HORNET MOTH**, feast on willow leaves.

The **WILLOW WARBLER**, a breeding migratory bird that comes here from Africa in summer, lives, nests and feeds in stands of willow along lakesides.

Alder

Not many Irish trees can put up with having their roots permanently in water, but alder is one that can. This tree is well able to grow in wet places, and its timber never rots. This why the bridges in Venice and the traditional clogs that Dutch people wear are made from alder wood.

Green algae

Green algae grow in the open water. These plants are so small that they can usually only be seen under a microscope. However, if lots of fertilisers are added to lake water, the algae grow like mad and the whole lake can turn green. This can happen if cattle slurry is sprayed on the surrounding fields to make the grass grow but is then washed off into the lake with the next shower of rain. Algae produce oxygen during the day because they use the sunlight to carry out photosynthesis. At night, however, when there is no sunlight, they do the opposite and use up the oxygen in the water to stay alive. Lots of green algae means that all the oxygen in the water gets used up at night – leaving none for fish and insects, who will then die. So waters full of green algae are very bad for biodiversity.

Edge plants

REEDS and **RUSHES** grow around the edges of the lakes where the water is shallower. The **COMMON REED** can move oxygen from its leaves down through its stem and out into the water through its roots, so it is very good for improving the amount of oxygen in the water.

Further out on the edge, with only their roots in the soil, is where purple water mint, yellow flag irises, blue forget-me-nots, white brook weed and green horsetails grow.

Animals

There is much less oxygen in water than in the air we breathe. Some freshwater creatures are very fussy and need a lot of dissolved oxygen to survive, others not much. If these fussy animals are common in lakes or rivers, this tells us that the water contains enough oxygen and so is very clean.

Fish

Fish get oxygen directly from water by taking it in through their mouths and letting it flow over their gills. Big clean lakes will have fish such as **TROUT** and **SALMON** living in them. They will also have smaller fish, such as **STICKLEBACKS**, **MINNOWS** and **STONE LOACHES**. Large, streamlined **PIKE** are there too. These are ruthless hunters and will grab and eat any small fish they can. Fast-flowing rivers and clean lakes also have a great variety of small creatures in the water. These range from water lice, water shrimps and leeches, which live under the water, to pond skaters and whirligig beetles that live on the surface of still ponds.

Freshwater pearl mussels

Freshwater pearl mussels can live up to 140 years, so they are easily Ireland's oldest animals. These are molluscs with two shells closed tightly together, and they get their food by filtering up to 50 litres of water every day through their **SIPHONS** and sieving out all the tiny creatures in it for food. But they need the cleanest possible water to live in – pristine waters – and Ireland has less and less of these every year.

Eighty per cent of all our freshwater pearl mussels live in just eight separate water bodies, which are specially protected for them.

Otters

The otter is our native wetland mammal. It lives in a holt, which is dug into the bank, and it feeds mainly on fish from the water. The **AMERICAN MINK** is an invasive pest that was introduced here in the 1950s. Not only does it eat fish, but it also eats the eggs and young of waterbirds and will even raid nearby henhouses and kill all the poor defenceless hens.

Water bats

DAUBENTON'S BAT – often called the water bat – hunts over rivers and lakes at night, feeding on flying insects such as moths, midges, mosquitoes and, on a very good night, a large mayfly swarm.

The Lough Ree monster

In 1960, three priests fishing on Lough Ree – one of the lakes on the River Shannon – saw a monster swimming not far from where they were. They described it as being over six metres long, swimming with its back above the water in a hump and with a head and neck half a metre long above the water at the front. They didn't get a photo – mobile phones hadn't been invented then. Other people crossing the lake have felt bumps and knocks against the bottoms of their boats. Nobody has ever got a photo of it ... yet.

Birds and flying insects

There are different habitats in watery areas with specific animals in each one. On the water will be the swimming waterbirds. In the air above the water is where the flying insects are. They are not feeding here but flying in great swarms looking for mates – sort of an insect disco, as it were.

Waterbirds that can swim

MALLARDS and **SWANS** are herbivores that feed on water plants. They can turn upside down and stretch their necks down to grab tasty plants growing underwater.

MOORHENS and **COOT** are large black waterbirds. Coot – the ones with the white bills – dive under the water to catch their food. They feed on snails, fish eggs and insects as well as water plants.

WATERHENS are smaller and have red bills with a yellow tip. They do not dive to catch their food, which includes insects, worms and fish, as well as spiders and berries, which they find on land.

BLACK-HEADED GULLS are often found inland feeding around lakes. They like fish but will eat carrion too – which is animals that are already dead.

Waterbirds that can't swim

Herons and kingfishers eat lots of fish, although neither of them can swim. The **GREY HERON** stands patiently in water waiting for a fish to swim past, which it then grabs and swallows head first.

The **KINGFISHER** watches from a tree branch, then leaps in when it spots a fish going by. It has to get out quickly with the fish before it drowns.

Dippers and grey wagtails feed on insects. The **DIPPER** catches caddis flies, stoneflies and mayflies by walking underwater, poking under stones.

The **GREY WAGTAIL** catches its food in mid-air by flying out from a perch.

Mayflies

Mayflies are insects that live in very clean waters, using the gills on their three back tails to get enough oxygen. They can live in the water for two or three years, but finally, when they have grown enough, they will develop wings. They then rise up out of the lake in great swarms – usually in May – and have one glorious party up in the air mating and laying eggs down into the water below. They drop dead after all this excitement. While they were around for two or three years living in the water, they only have one day as fully grown adults.

Dragonflies

Dragonflies have the same sort of life cycle, but they manage to live much longer as adults. They don't swarm but emerge singly and fly around in the air, catching small insects and looking for mates.

A lake with lots of flying dragonflies and their smaller relatives – the damselflies – will have lots of other biodiversity too.

Remember, these plants and animals are much rarer than others – how many can you find?

Osprey – the fish hawk

Ospreys are fish-eating birds of prey. They build their nest on the top of a large tree near a lake – using the same tree every year. They feed almost entirely on fish, which they catch by either diving in head-and-feet first and grabbing it underwater, or by flying just above the water and grabbing it with their strong talons. They winter in Africa and breed in northern Europe. Because they had been extinct in Ireland since the end of the 1700s, nine chicks from Norway were released into the wild here in 2023. That same year, a pair had naturally bred in County Fermanagh and raised two chicks all on their own. Look out for them in summertime.

Natterjack toad

Natterjack toads are our only native species of toad. They breed in freshwater ponds on the Dingle peninsula during the month of May, having been in hibernation from November till April. Unlike frogs, which produce clumps of eggs, toads lay their eggs in strings 1.5 metres long, which they wind around plants in the pond.

They create a terrible racket during the breeding season – especially at night. You will definitely hear them at this time, so May is the month to look for them. These small brown toads with a yellow stripe down their back cannot jump like frogs. No species of toad can. But these – being Kerry toads – run rather than walk as they chase insects and spiders to eat.

White-tailed sea eagle

This is our biggest bird of prey, with a wingspan of 240 centimetres, which is absolutely huge. These eagles hunt for fish on big lakes in the south and west of Ireland. They are **SCAVENGERS** too, eating any dead animals they find as well as stealing food from otters, gulls and other seabirds. They need 600 grams of meat every day, so finding food is a full-time job. They were made extinct in Ireland in 1898 – persecuted to death by gamekeepers at that time. Reintroduced from Norway in 2007, they are now thriving again and breeding here every year. They mate for life and can rear two chicks a year if food is plentiful.

Diving bell spider

This is the only spider in Ireland, or indeed in the world, that can live underwater in lakes and ponds. It spins a web between underwater plants, then brings air down underwater on its body hairs and releases it in the web, making a big air bubble. It breathes this air then rushes out to catch flies and even tadpoles.

It has ferocious fangs and can stick them into its prey, which it quickly kills with its poisonous venom. The air in the body hairs makes the spider look silvery as you bring it to the surface in your pond net. Be sure to put it back carefully afterwards!

THE GREAT IRISH NATURE CHALLENGE

ON THIS MAP, you will find a selection of places to visit around Ireland. Each one of these locations is an example of the habitats featured in this book – woodlands, bogland, the seaside, gardens and parks, grassland and rivers and lakes. Some are places you could go to look for the creatures mentioned in The Great Irish Nature Challenge – like the natterjack toads, which can be found on Dingle Peninsula, or the goosander, which lives in the oak woodlands in Glendalough. Others you might visit to walk through a specific habitat, like a seaside walk on Bull Island. And indeed, you must be sure to visit the National Museum of Ireland to see what has been unearthed in Irish bogs!

WOODLANDS

- Charleville Forest, Co. Offaly
- Glendalough, Co. Wicklow
- Glengariff, Co. Cork
- Gosford Forest Park, Co. Armagh
- Jenkinstown Wood, Co. Kilkenny
- Killarney National Park, Co. Kerry
- Oldhead Wood, Co. Mayo

GARDENS AND PARKS

- Connemara National Park, Co. Galway
- Dalgan Park, Co. Meath
- Florence Court Gardens Co Fermanagh
- Mount Congreve Gardens, Co. Waterford
- Mount Stewart Gardens, Co. Down

RIVERS AND LAKES

- Barrow River, Co. Carlow
- Lough Beg, Co. Derry
- Lough Ree, Co. Longford
- Pollardstown Fen, Co. Kildare
- River Shannon
- Rogerstown Estuary, Co. Dublin
- Western Shore of Lough Neagh,
- Co. Tyrone
- Wexford Wildfowl Reserve, Co. Wexford

BOGLANDS

- Abbeyleix Bog Project, Co. Laois
- Cashel Bog, Co. Laois
- Coggalbeg, Co. Roscommon
- Eshbrack Blanket Bog, Sliabh Beagh, Co. Monaghan
- Faddan More Bog, Co. Tipperary
- Knockdrin Bog, Co. Offaly
- Littleton Bog, Co. Tipperary
- National Museum of Ireland, Dublin
- Scragh Bog Trail Head, Co. Westmeath

THE SEASIDE

- Baltray Dunes, Co. Louth
- Bull Island, Co. Dublin
- Dingle Peninsula, Co. Kerry
- Great Skellig, Co. Kerry
- Rathlin Island, Co. Antrim
- Streedagh Point Sand Dunes, Co. Sligo

GRASSLANDS

- Burren National Park, Co. Clare
- Curragh, Co. Kildare
- Glenade Cliffs, Co. Leitrim
- Sheskinmore, Co. Donegal
- Tory Island, Co. Donegal

TORY ISLAND
DONEGAL
RATHLIN ISLAND
ANTRIM
SHESHKINMORE
DONEGAL
LOUGH BEG
DERRY
MOUNT STEWART GARDENS
DOWN
LOUGH NEAGH (WESTERN SHORE)
TYRONE
STREEDAGH POINT SAND DUNES
SLIGO
GLENADE CLIFFS
LEITRIM
ESHBRACK BOG
MONAGHAN
GOSFORD FOREST PARK
ARMAGH
FLORENCE COURT
FERMANAGH
OLDHEAD WOODS
MAYO
COGGALBEG BOG
ROSCOMMON
BALTRAY BEACH
LOUTH
CONNEMARA NATIONAL PARK
GALWAY
SCRAGH BOG TRAIL HEAD
WESTMEATH
DALGAN PARK
MEATH
LOUGH REE
LONGFORD
KNOCKDRIN BOG
OFFALY
NATIONAL MUSEUM OF IRELAND
DUBLIN
BULL ISLAND
DUBLIN
CHARLEVILLE FOREST
OFFALY
POLLARDSTOWN FEN
KILDARE
RIVER BARROW
CARLOW
CURRAGH
KILDARE
GLENDALOUGH
WICKLOW
BURREN NATIONAL PARK
CLARE
FADDAN MORE BOG
TIPPERARY
CASHEL BOG
LAOIS
ABBEYLEIX BOG PROJECT
LAOIS
JENKINSTOWN WOOD
KILKENNY
LITTLETON BOG
TIPPERARY
MT. CONGREVE GARDENS
WATERFORD
DINGLE PENINSULA
KERRY
KILLARNEY NATIONAL PARK
KERRY
GREAT SKELLIG
KERRY
GLENGARIFF
CORK

Acknowledgements

Éanna:

I WOULD LIKE to thank the team at Gill for all their hard work. Sarah Liddy, who suggested that I should write this book and refused to take no for an answer. Aoife Harrison, who put order on what I had written so that I didn't repeat myself. Aoibheann Molumby, Charlie Lawlor, Fiona Murphy, Kristen Olson, Graham Thew, Liz White and Georgina Cope were also part of the team that worked so tirelessly to get the book over the line in time. Barry Falls has been a wonderful and very patient illustrator – his drawings are great and also scientifically accurate. And, most of all, my patient husband John Harding, who fears, with some justification, that I shall never retire.

Barry:

THIS BOOK WAS a labour of love – and a real opportunity for me to spend many quality hours researching, and marvelling at, the wondrous biodiversity that we are lucky to have inherited on our doorsteps in this special corner of the world. It is incredible. Thanks to Éanna for being my guide, and to the great team at Gill for bringing us together. I'll dedicate this book to my children, Jack, Noah and May, and to all the children in Ireland who, like me, are endlessly curious about the natural world, and captivated by its fragile beauty. They are its future.

Index

Gill Books

Hume Avenue

Park West

Dublin 12

www.gillbooks.ie

Gill Books is an imprint of M.H. Gill and Co.

978 18045 8055 4

Design origination by
www.grahamthew.com

Designed by
Liz White Designs

Edited by
Emma Dunne

Proofread by
Bríd Nowlan

Index by
Eileen O'Neill

Printed and bound by
Printer Trento, Italy

This book is typeset in Sofia Pro, Bookmania and Venti CF.

The paper used in this book comes from the
wood pulp of sustainably managed forests.

A CIP catalogue record for this book is
available from the British Library.

5 4 3 2 1

FSC
www.fsc.org
MIX
Paper | Supporting
responsible forestry
FSC® C015829

COLLECT THEM ALL

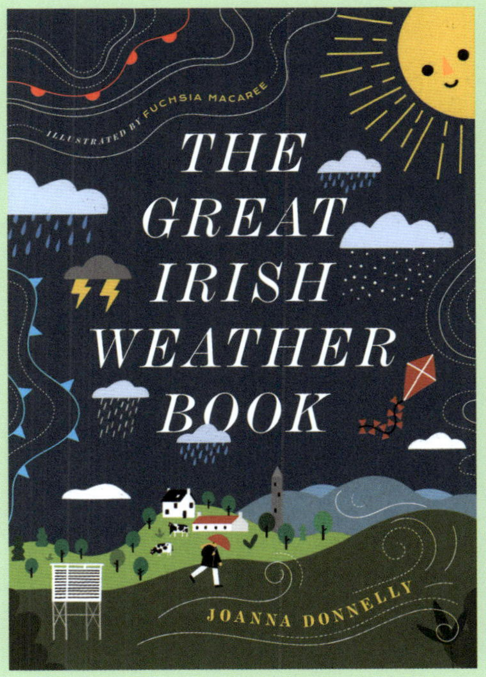

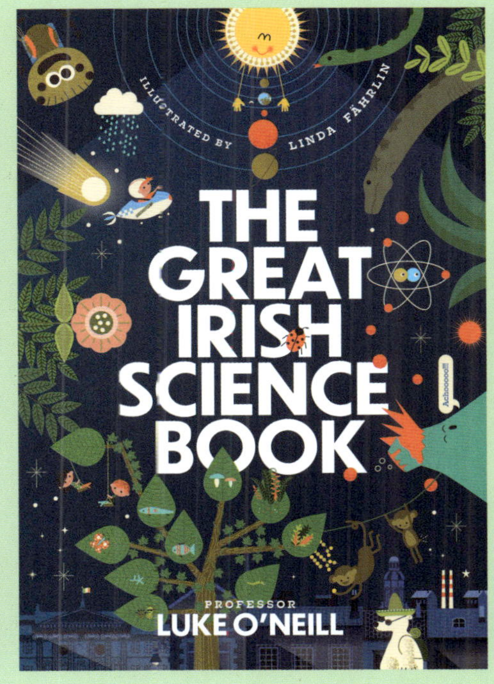

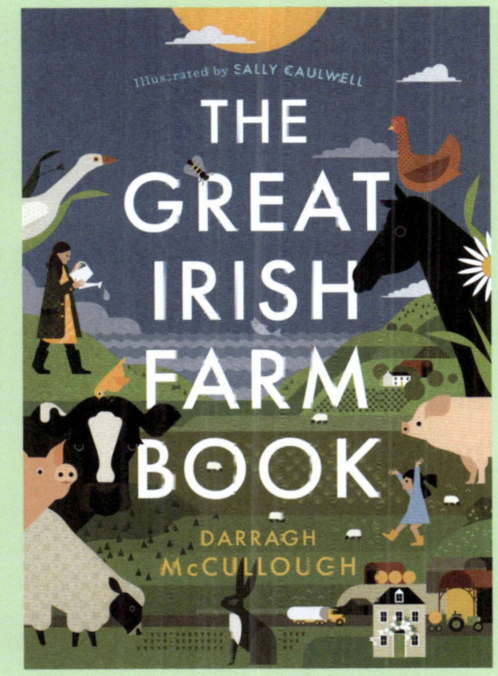

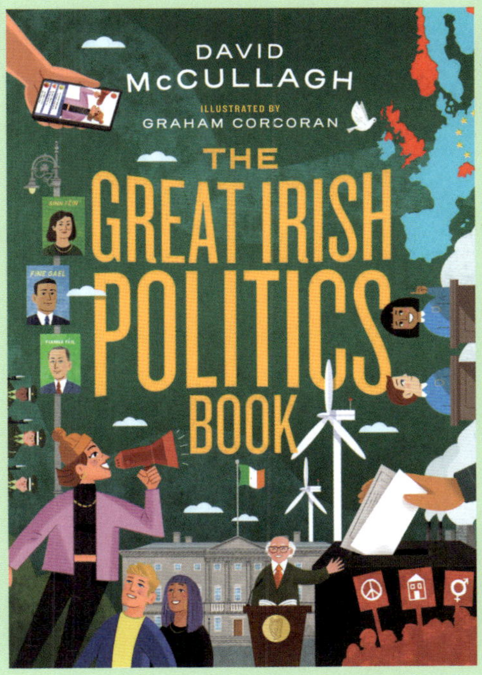

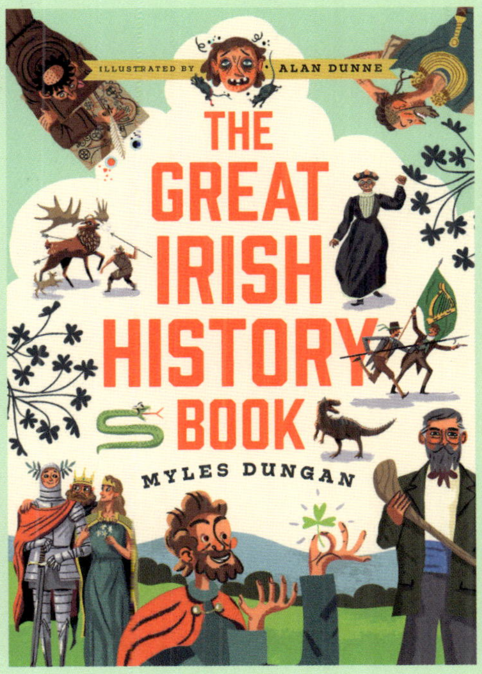

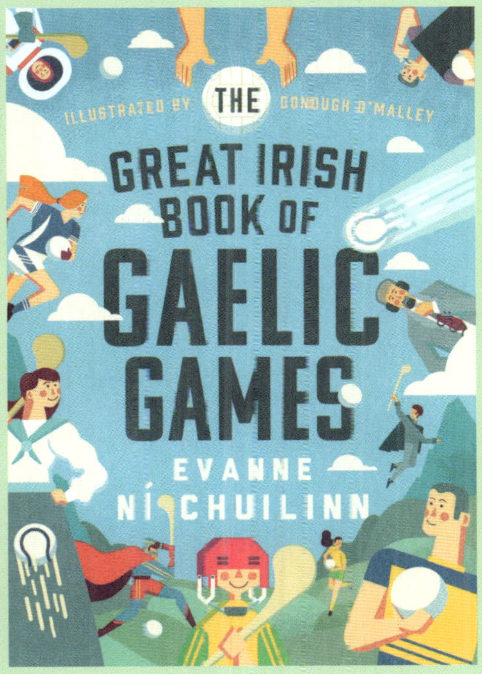